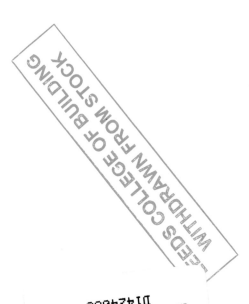

Illustrated Building Pocket Book

For my parents
with gratitude and love

Illustrated Building Pocket Book

Second Edition

Roxanna McDonald

ELSEVIER

AMSTERDAM • BOSTON • HEIDELBERG • LONDON •
OXFORD • NEW YORK • PARIS • SAN DIEGO •
SAN FRANCISCO • SINGAPORE • SYDNEY • TOKYO
Butterworth-Heinemann is an imprint of Elsevier

B H

Butterworth-Heinemann is an imprint of Elsevier Ltd
Linacre House, Jordan Hill, Oxford OX2 8DP
30 Corporate Road, Burlington, MA 01803

First edition 1999
Reprinted 2000, 2001, 2003, 2004
Second edition 2007

British Library Cataloguing in Publication Data
A catalogue record for this book is available from the British Library

Library of Congress Cataloguing in Publication Data
A catalogue record for this book is available from the Library of Congress

ISBN 13: 978-0-75-068015-8
ISBN 10: 0-75-068015-6

For information on all Elsevier Butterworth-Heinemann publications visit our website at www.books.elsevier.com

Typeset by Cepha Ltd
Printed and bound in Great Britain

07 08 09 10 11 10 9 8 7 6 5 4 3 2 1

Contents

Superstructure – internal walls

Superstructure – roofs

Superstructure – stairs

Superstructure – chimneys

Superstructure – floors

Superstructure – openings in walls

Superstructure – fixings

Finishes

Glazing

Services – drainage and plumbing

Services – electrical

Foreword

Since this book was published six years ago under its original title '*An Illustrated Building Glossary*' changes in the priorities of approach to design and management have accelerated, and an understanding of the need for broadly-educated professionals with an understanding of their interdisciplinary responsibilities is increasingly accepted. The areas in which the book has been extended witness this change.

Certainly, concern for the protection of the environment has been on the agenda of the specialist for twenty years but the public have finally accepted the seriousness of the situation we have made for ourselves from the profligate use of our natural inheritance. It is now recognised that only around a quarter to a third of the energy and resource expended in buildings with a fifty year life span is used in their construction. Consequently, the whole life cycle of a building becomes a critical consideration from the inception of a project. To achieve an environmentally sustainable building environment, every professional must have an understanding of the development process if they are to be able to contribute to it in a balanced way. The computer has presented a way not only of drawing our ideas but also of coordinating the process of design, of anticipating a building's performance and monitoring the results. The computer is now a central tool in procurement as design is increasingly moved directly from the computer model to smart manufacture.

In 1966, when I joined Arup Associates at No. 8, Fitzroy Street, London W1, a basement wall some twenty-five feet long was required to crunch the numbers, hour after hour, in Ove Arup and partners' design of the Sydney Opera house shells; one is reminded of scenes in the epic silent film 'Metropolis'. I now sit in front of a laptop with vastly more potential and, if I were clever enough, I could swallow the problem in one bite! This revolution now allows ideas to be shared and tested from their myriad facets and, from it, we are able to optimise solutions within a new four-dimensional freedom.

But what further updates will be needed to the *Illustrated Building Pocket Book* in ten years time? Here optimism clashes with pessimism as we are in the hands of politicians who may well push in the contrary direction to the needs of 'Everyman'. Maybe I can be permitted to speculate about changes to any future edition.

An accelerating move away from a carbon-based economy towards sustainable sources will sharpen our need for renewables: This change will materialise in a number of ways but most dramatically, as the price tumbles, the use of photovoltaics will become universal and will require ingenious transformations to our designs. As the impact of a building's life cycle permeates the professions, not only will elements within buildings be chosen to minimise maintenance but also, increasingly the house-buying public will start to judge a potential purchase in terms of the likely life-cycle running costs as well as the capital investment. Lastly, but sadly decades away, when the penny finally drops, that endless economic growth will eventually lead to our demise rather than our redemption; there will be a massive change which will touch every part of our build environment and our means of achieving it.

Back to the present content of the Pocket book: having been both a practitioner and a university teacher, I am acutely aware of the tendency to treat the young student and the practitioner as different beings. Of course, the student's knowledge will be less rounded, but the act of learning is a continuing process, and to revisit words describing hard fact, presented in a new and illuminating way, is to be in a position not only to re-evaluate those facts but also to explore the ideas that stem from them.

The range of knowledge and skill required to operate in architecture and construction is immense, and in studying one page of the new pocket book, this is all too obvious. Each *word* related to an item in an illustration is simply a flag marking the tip of an iceberg. Each carefully chosen word is filled with potential; it introduces one piece, one aspect, of one element of the process of building. In turn, the process of building is but the beginning of defining the place for a society to function; a place where the buildings gain a symbolic presence. So the volume you are holding is a book with many resonances.

The new volume may have changed its title but it is still a glossary and a very unusual one. Here, refreshingly, the explanation is visual, and through the clarity and completeness of the 'visual paragraph', meaning is given and a context described, in a form normally thought to require words. Indeed, the only piece of written text by the author is her five hundred word preface: an admirable achievement.

What of the presentation of this book? The structure and presentation themselves are worth study as a piece of design. A large quantity of material has been explored, digested and synthesised to present a core of information clearly without becoming simplistic. Each illustration is from the same hand and must have taken hours to draw, never mind the weeks of research, assimilation and evaluation involved. Anyone who has tussled with a small design problem and who has attempted to present a solution in a simple line drawing is all too aware of the time taken in graphical study before undertaking the drawing itself.

A book of this kind is only useful if it presents the possibility of relating a term to a subject area and referring from there to an in-depth bibliography. The author's bibliography is short but all the reference books are in common use, and from those, further channels can be explored.

In summary, the value of such a book for the experienced professional or crafts person is that it has much good straightforward information about the processes themselves, communicated in an attractive way. It contains much for the student to learn; and for the experienced, it contains much that we once knew and are ashamed to admit we have forgotten!

Richard Frewer
Director of Arup Associates 1977-2001
Chair Professor of Architecture, University of Bath 1991-2000
Chair Professor of Architecture, University of Hong Kong 2000-2005

Preface

It is not the intention of this book to provide an exhaustive list of building terms or to attempt a comprehensive teaching of building technology. There are many specialist encyclopedias, dictionaries and construction manuals which supply ample information in this respect. The book sets out to be primarily a *communication tool* using the *visual reference* as vocabulary.

The creation of a building is the result of a complex process of interaction between people of different professions, views, even nationalities, with varying technical knowledge and motivation. Architects, who at the centre of it all, often find themselves as 'interpreters' between the participants, use image as the safest interface.

The language we each use grows from our own personal experience and, sometimes, the same word can mean different things to different people depending on the circumstances in which they have learnt it. The same can apply to building terms.

Images on the other hand leave little room for ambiguity, and many a time a site query or dispute has been sorted out with the aid of a sketch scribbled on a wall! Words express ideas we have of tangible objects and can be classified into a system such as an alphabetical dictionary or be placed in context as in a thesaurus. The same can apply to images – they can be attached to words arranged in alphabetical order or they can be placed in the context to which they are relevant.

It is the later system this book has adopted, attempting to present the terms in the context in which they are likely to apply. The main building terms that form the language of construction are set out to follow the logical sequence of the building process. If one can't remember the right word or wants to know what a specific part is called, it should be simple enough to locate it on the sketch in the relevant section. Similarly, by placing something visually in context it should be much easier to learn terms rather than to memorize their abstract definition. At the same time, the index permits the reverse to take place making it possible to find the context of a given word.

The drawings are simple line sketches concerned mostly with descriptive clarity rather than comprehensive accuracy. The diagrams are intended to identify the sequence and relationships as well as particular terminology.

Compiled primarily as a visual checklist for students and early stages of practice building professionals, the book is also meant to help communication with the other participants to the building industry.

Its spirit, I hope, echoes the intentions of a much older introduction from which I quote below as it is as valid today as it was when it was first written.

> IT is uſeful Knowledge only, that makes one Man more valuable than another, and eſpecially that part of Knowledge, which immediately concerns the Buſineſs he is to live by; and therefore, if this Work ſhould prove a Help to the Improvement of Knowledge in *Youth*, (for whoſe Sakes 'tis chiefly intended:) and be no Affront to the *ſage Workman*, by re-informing him of thoſe Rules which have ſlipt his Memory, and informing him of others which he never knew, it will anſwer the deſired End of their hearty Well-wiſher,
>
> *London, March 25th*, 1741. THO. LANGLEY.

From the introduction to:

THE

BUILDER's JEWEL:

OR, THE

YOUTH's INSTRUCTOR,

AND

WORKMAN's REMEMBRANCER.

EXPLAINING

SHORT and EASY RULES,

Made familiar to the meaneſt Capacity,

For DRAWING and WORKING.

By B. and T. LANGLEY.

LONDON:

Printed for *R. Ware*, at the *Bible* and *Sun* in *Amen-Corner*, near *Pater-Noſter-Row*. MDCCXLI. [Price 4s. 6d.]

Acknowledgements

I am grateful to the following people and organisations for their supportive help during the preparation of the first edition of this book:

Rob Dark, Architect, UK
B. Goilav, Structural Engineer, France
Dan S. Hanganu, Architect, Montreal, Canada
Claude and Anca Lemaire, Architects, France
Biblioteque Centre Pompidou, Paris, France
The RIBA Library, London, UK
Veronique Thierry, Isabelle Mathieu, Monique Beranger, Architects, Paris, France
Beatrice Jubien, France
Special thanks to Jane Fawcett whose generous advice and personal example were an inspiration.

My further thanks for help in preparing this edition to:

Dominic Hailey (CAD director) of WORK OR PLAY for his advice on the computer drawing chapter. A London based multi-disciplinary organisation, WORK OR PLAY (www.workorplay.org) specialise in CAD data management systems for the construction industry advising and training architects in the UK and Europe.

Jason Dunn DSc (Hons) MB Eng Tech RICS MAIBS – for his help in updating the Construction Control and Building Control in London chapters.

Finally, my most grateful thanks to my editor Alex Hollingsworth for his help, advice and staunch support in completing this new edition.

About the Author

Roxanna McDonald is a practising architect. She works in the UK, France and Eastern Europe advising on a wide range of professional issues ranging from the conservation of historic buildings to the re-building of infrastructure damaged by natural and man-made disasters and building-related environmental issues.

Other books written by Roxanna McDonald:

The Fireplace Book - Architectural Press 1984
Illustrated Building Glossary - Butterworth Heineman 1999
Introduction to Natural and Man Made Disasters and their effects on Buildings - Architectural Press 2003

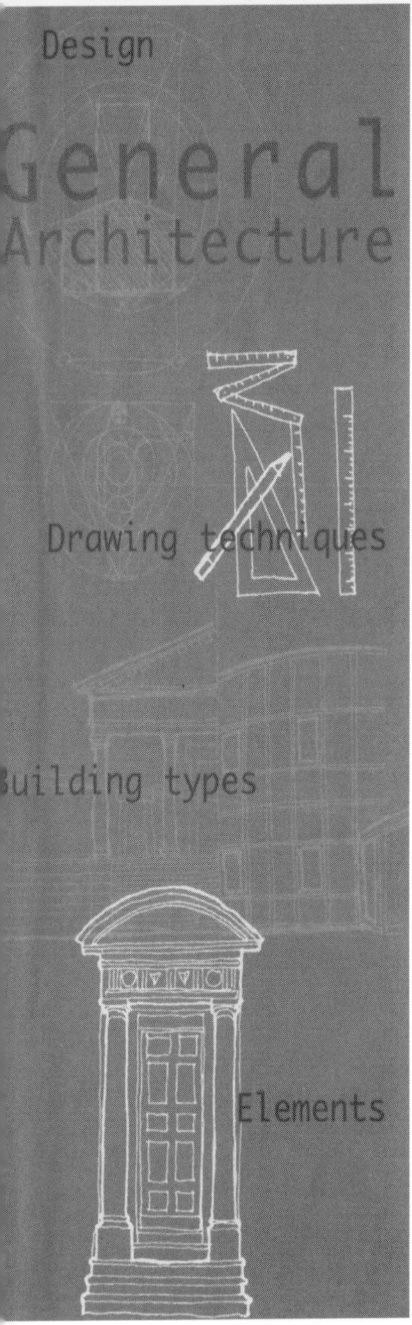

I. General Architecture

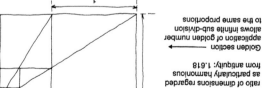

Golden number:
ratio of dimensions regarded as particularly harmonious from antiquity: 1.618

Golden section →
application of golden number allows infinite sub-division to the same proportions

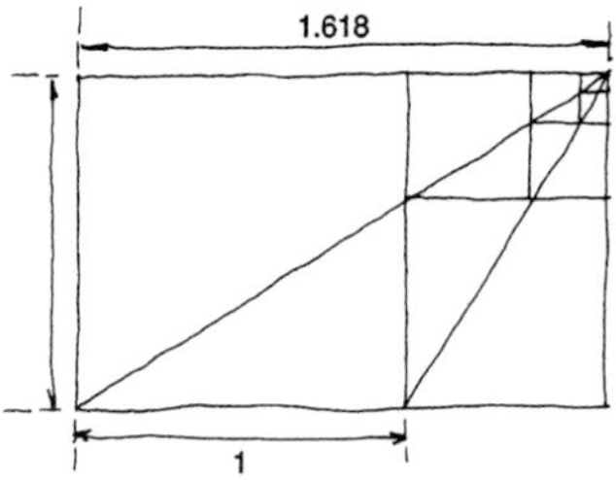

Modulor; Le Corbusier's application of the golden number proportions to human dimensions

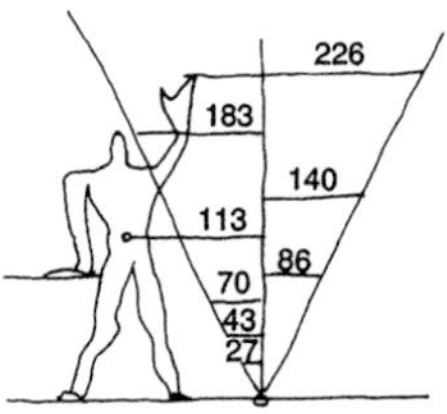

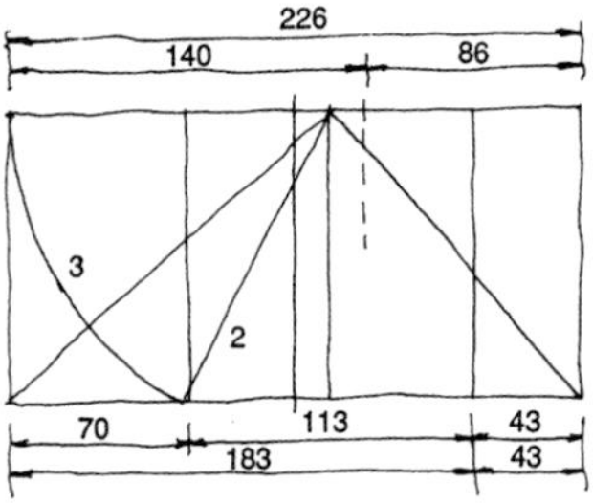

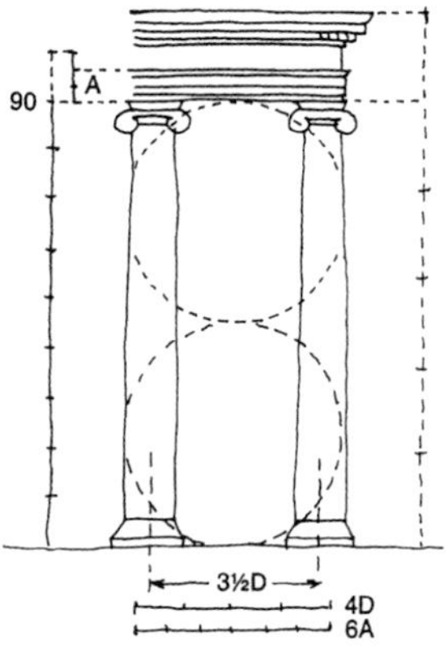

Classical orders (Vitruvius)

Module = half diameter of column base

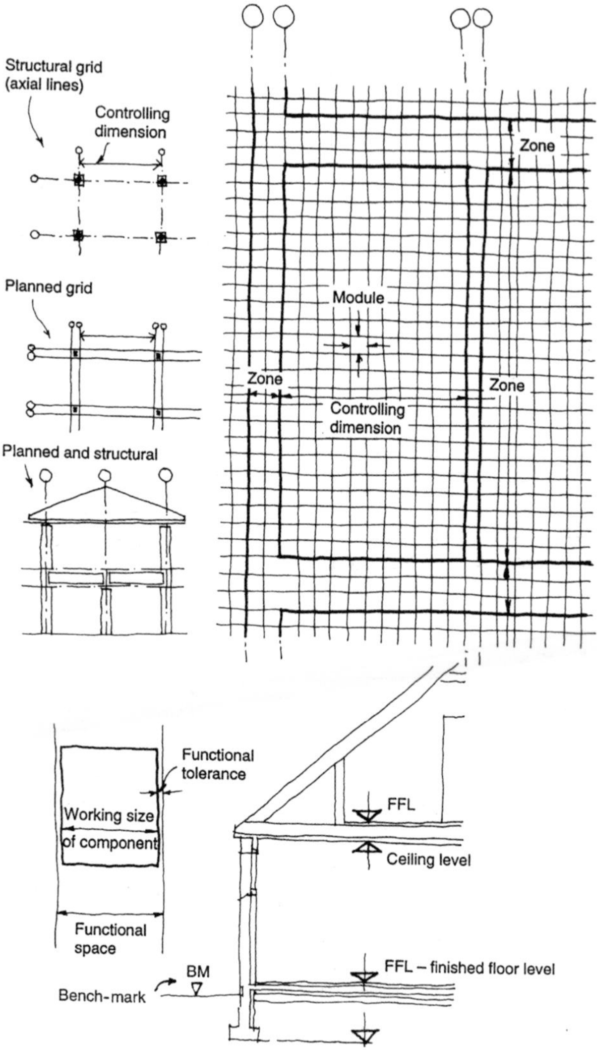
Structural grid
(axial lines)
Controlling
dimension
Planned grid
Planned and structural
Zone
Module
Zone
Zone
Controlling
dimension
Functional
tolerance
Working size
of component
Functional
space
BM
Bench-mark
FFL
Ceiling level
FFL – finished floor level

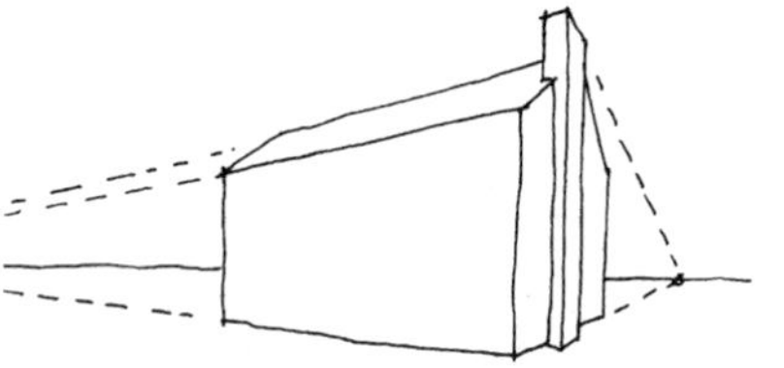

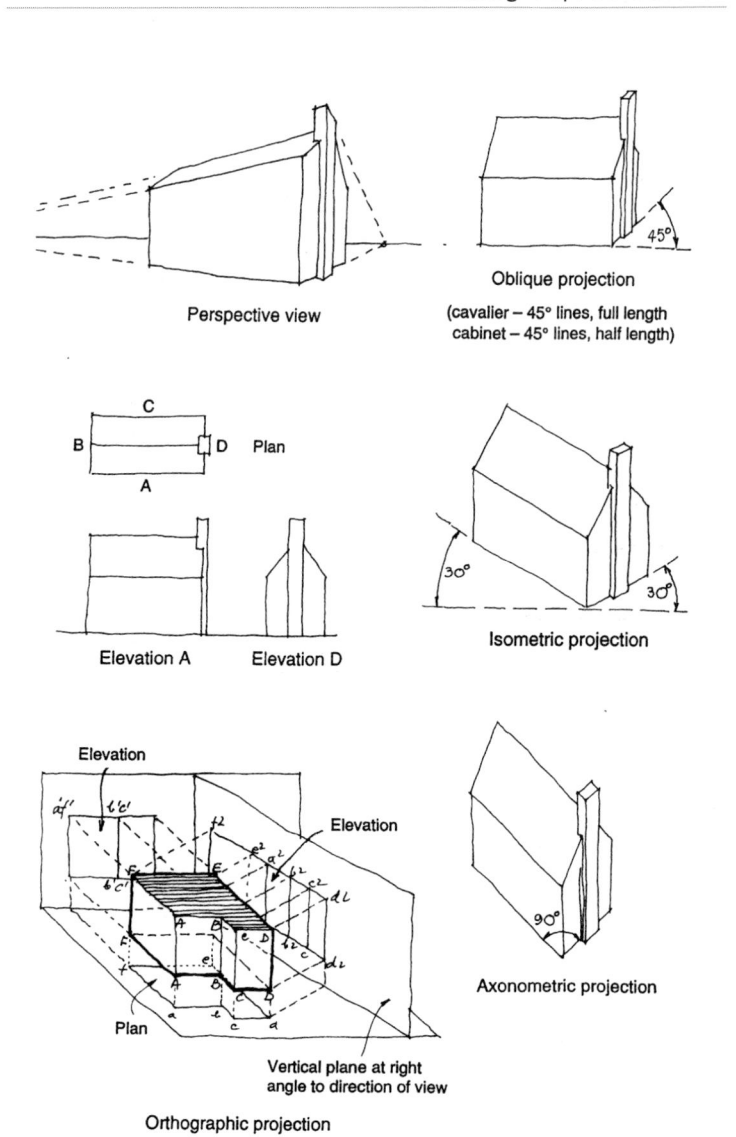

Perspective view

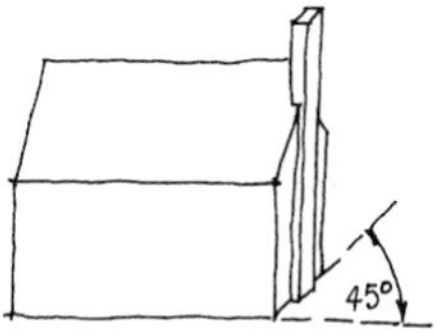

Oblique projection

(cavalier – 45° lines, full length
cabinet – 45° lines, half length)

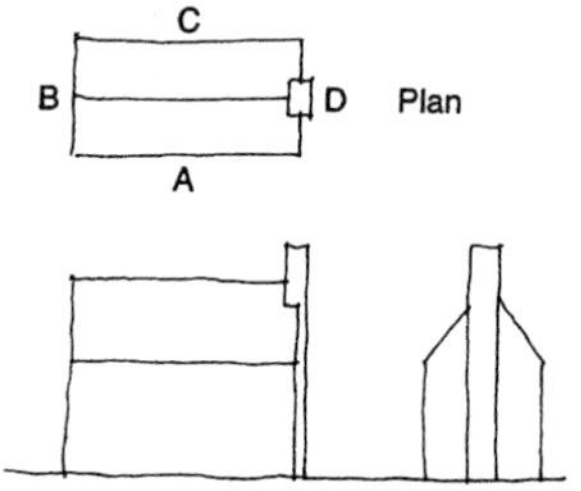

Elevation A

Elevation D

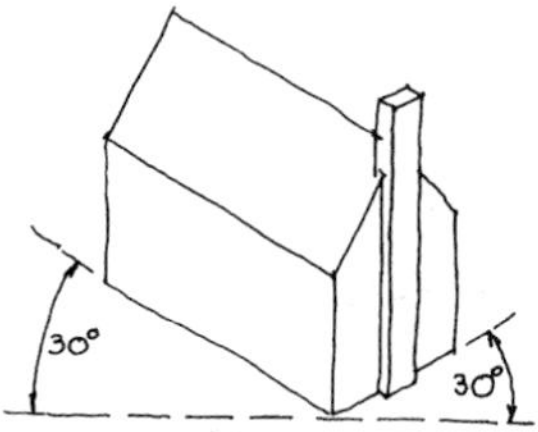

Isometric projection

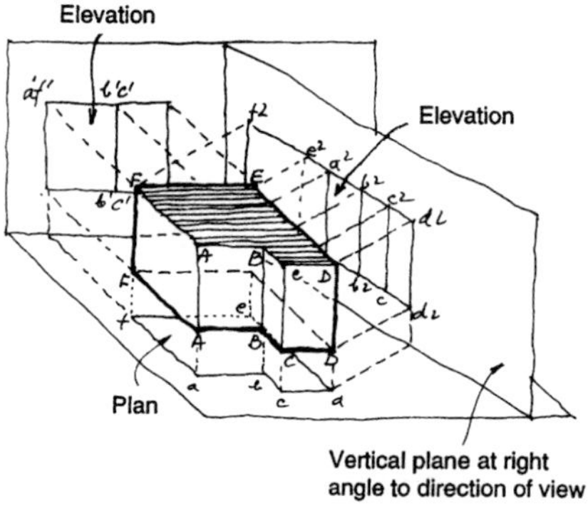

Orthographic projection

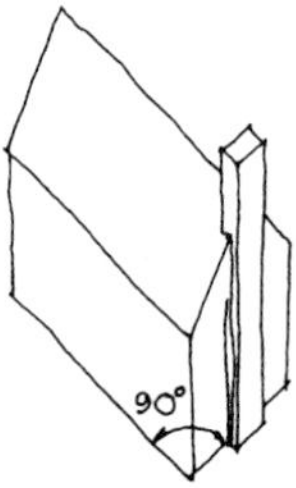

Axonometric projection

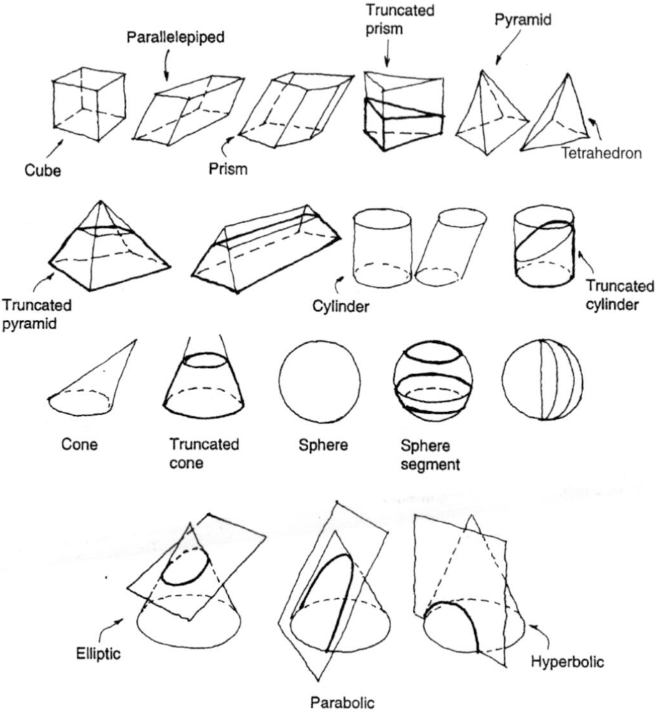
Parallelepiped
Truncated prism
Pyramid
Cube
Prism
Tetrahedron
Truncated pyramid
Cylinder
Truncated cylinder
Cone
Truncated cone
Sphere
Sphere segment
Elliptic
Parabolic
Hyperbolic

A size	mm
A0	841x1189
A1	594x841
A2	420x594
A3	297x420
A4	210x297
A5	148x210
A6	105x148
A7	74x105
A8	52x74
A9	37x52
A10	26x37

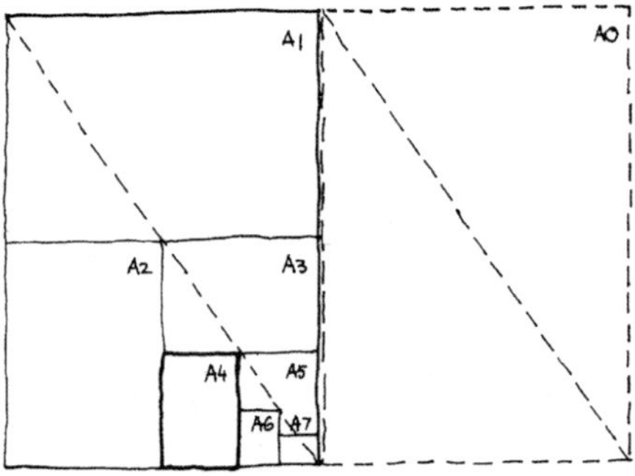

A-sizes

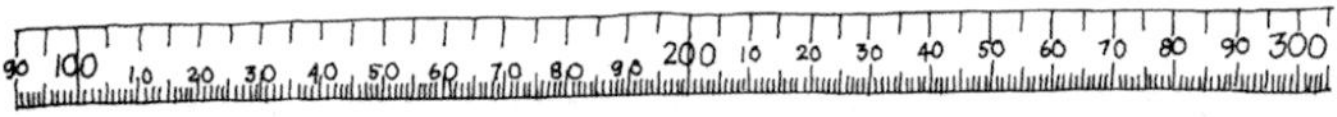

Metric graduation markings

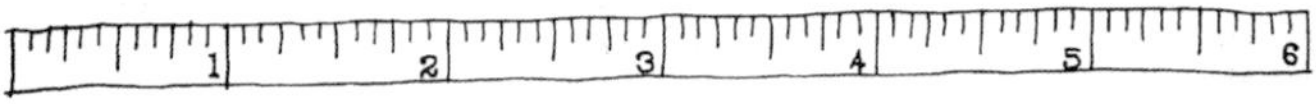

Imperial graduation markings

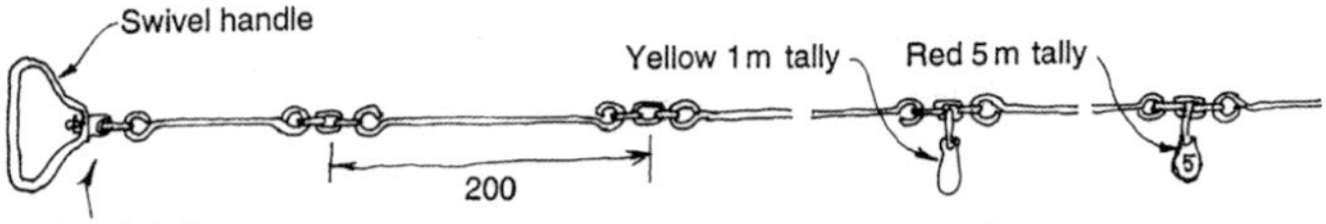

Dimensional coordination

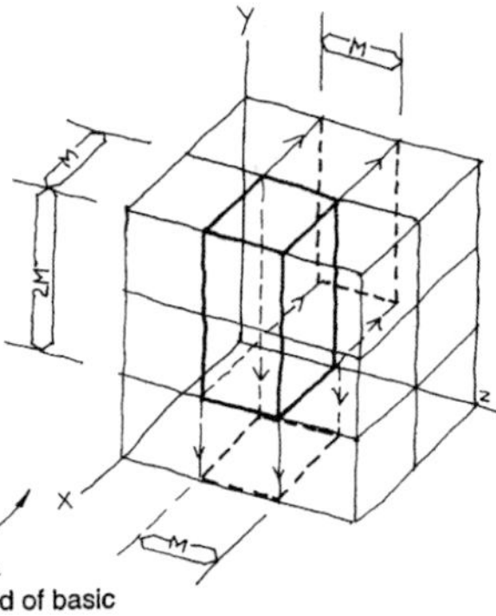

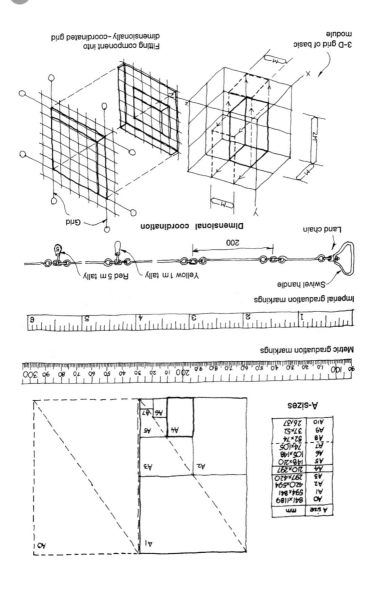

3-D grid of basic module

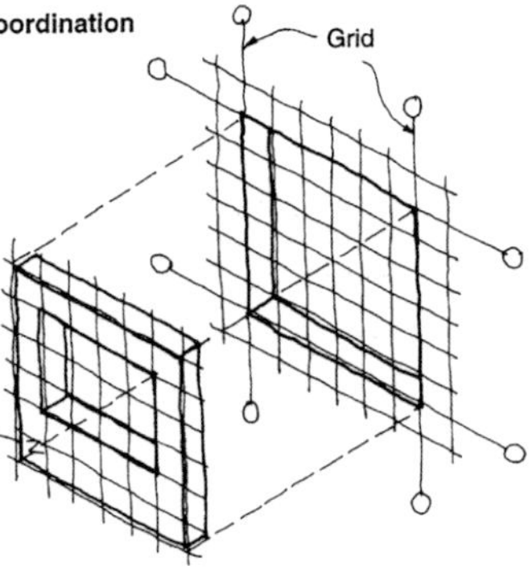

Fitting component into dimensionally-coordinated grid

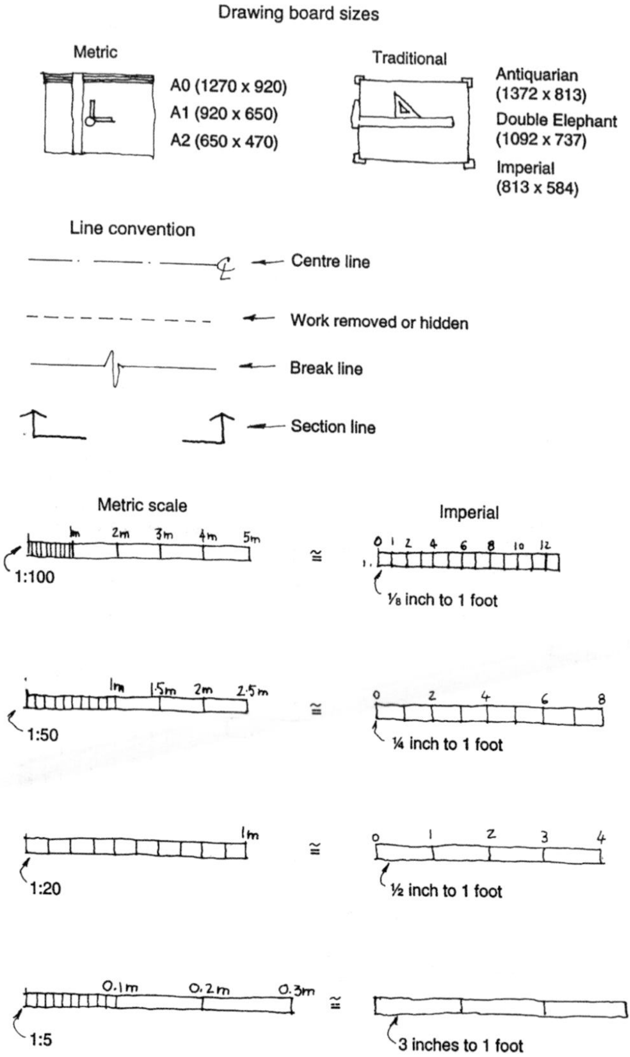
Drawing board sizes
Metric
A0 (1270 x 920)
A1 (920 x 650)
A2 (650 x 470)
Traditional
Antiquarian (1372 x 813)
Double Elephant (1092 x 737)
Imperial (813 x 584)
Line convention
Centre line
Work removed or hidden
Break line
Section line
Metric scale
Imperial
1m 2m 3m 4m 5m
0 1 2 4 6 8 10 12
1:100
⅛ inch to 1 foot
1m 1.5m 2m 2.5m
0 2 4 6 8
1:50
¼ inch to 1 foot
1m
0 1 2 3 4
1:20
½ inch to 1 foot
0.1m 0.2m 0.3m
1:5
3 inches to 1 foot

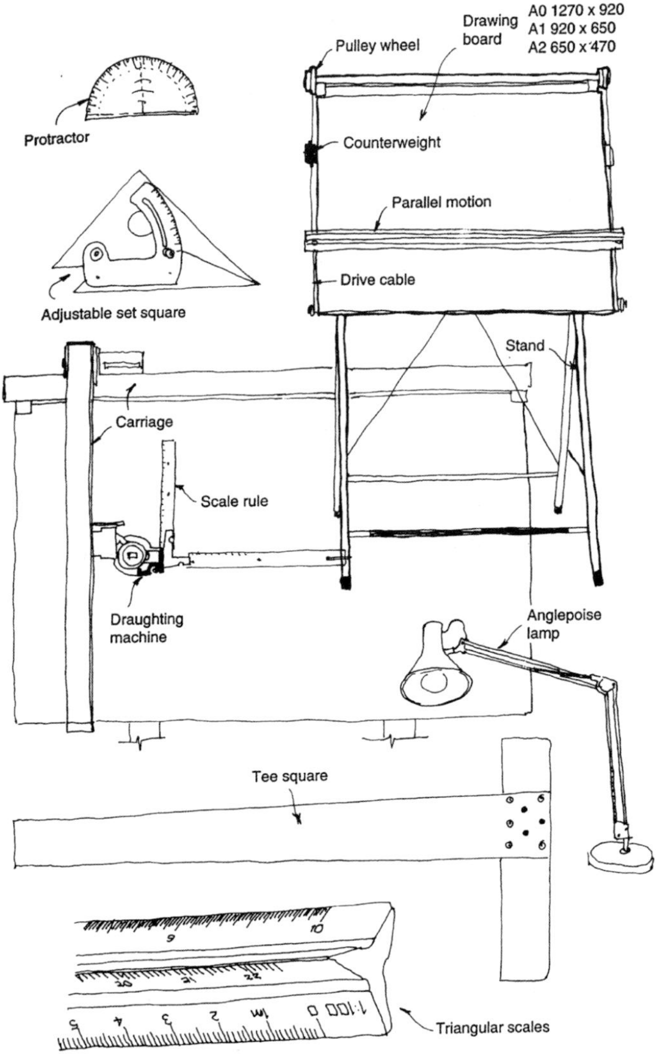

A0 1270 x 920
A1 920 x 650
A2 650 x 470
Drawing board
Pulley wheel
Protractor
Counterweight
Parallel motion
Adjustable set square
Drive cable
Stand
Carriage
Scale rule
Draughting machine
Anglepoise lamp
Tee square
1:100 0 1m 2 3 4 5
Triangular scales

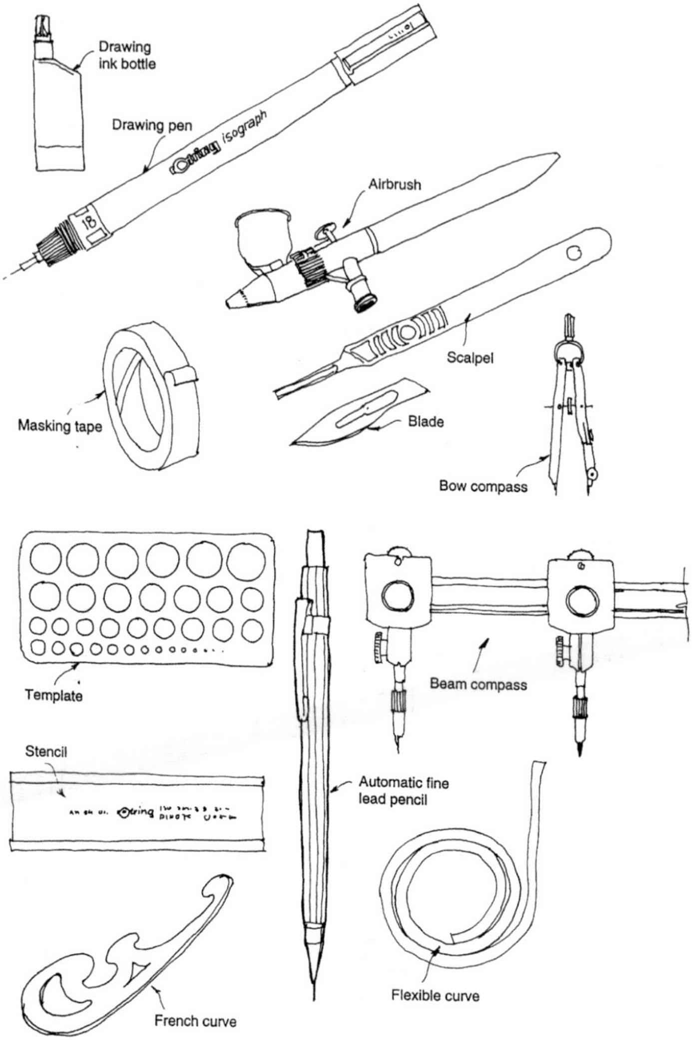
Drawing
ink bottle
Drawing pen
isograph
Airbrush
Masking tape
Scalpel
Blade
Bow compass
Template
Beam compass
Stencil
Automatic fine
lead pencil
French curve
Flexible curve

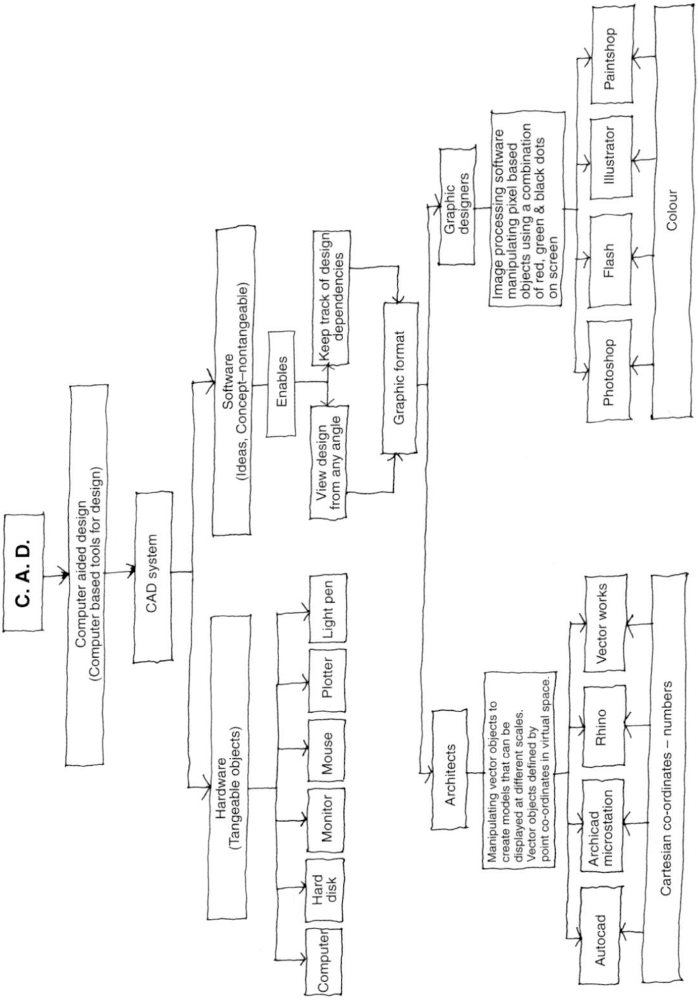
C. A. D.
Computer aided design
(Computer based tools for design)
CAD system
Hardware
(Tangeable objects)
Computer
Hard disk
Monitor
Mouse
Plotter
Light pen
Software
(Ideas, Concept–nontangeable)
Enables
View design from any angle
Keep track of design dependencies
Graphic format
Architects
Manipulating vector objects to create models that can be displayed at different scales. Vector objects defined by point co-ordinates in virtual space.
Autocad
Archicad microstation
Rhino
Vector works
Cartesian co-ordinates – numbers
Graphic designers
Image processing software manipulating pixel based objects using a combination of red, green & black dots on screen
Photoshop
Flash
Illustrator
Paintshop
Colour

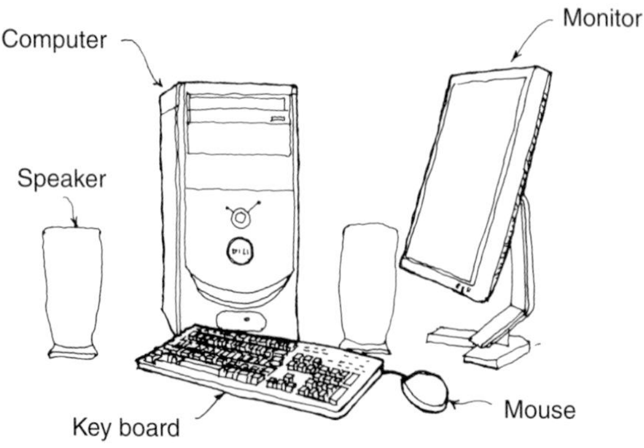
Computer
Monitor
Speaker
Key board
Mouse

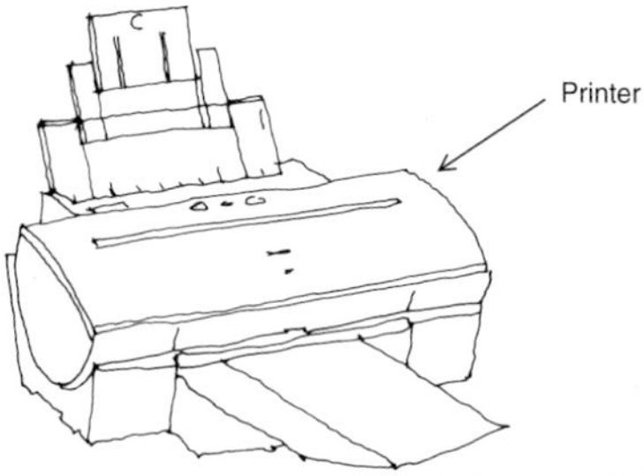
Printer

Plotter

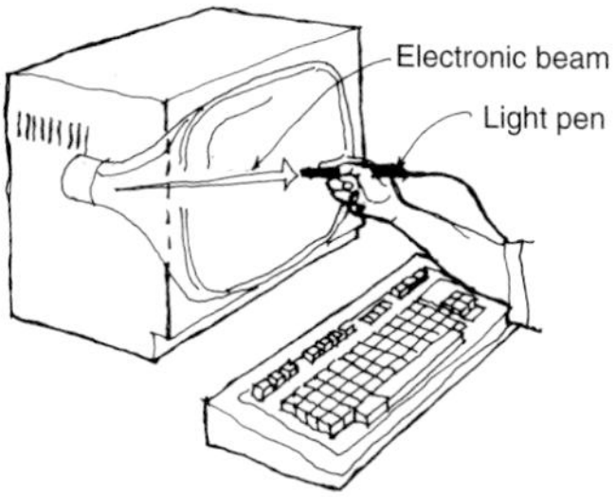
Electronic beam
Light pen

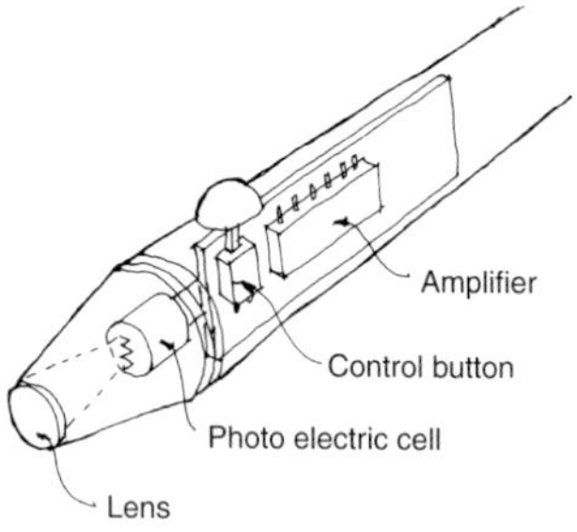
Amplifier
Control button
Photo electric cell
Lens

AUTOCAD		VECTOR WORKS
Model space	=	Design layer
Paper space	=	Sheet layer
Layer	=	Class
X ref	=	Working group reference
UCS	=	Origin
Block	=	Symbol
Point	=	Locus
Properties palette	=	Object info palette
Design centre	=	Resource browser

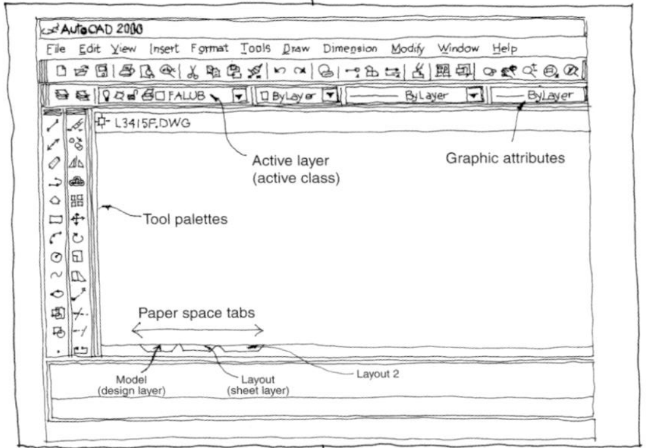

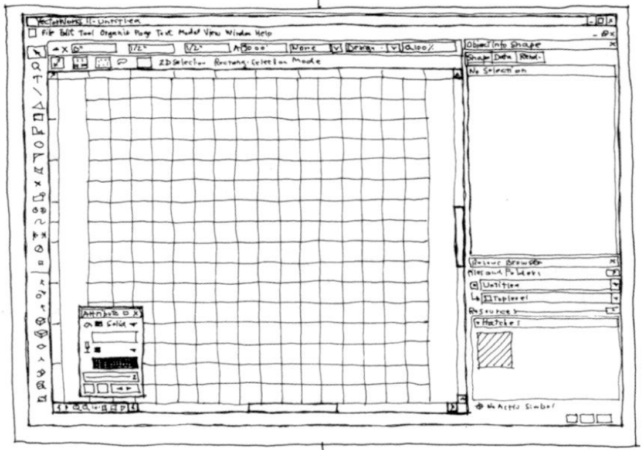

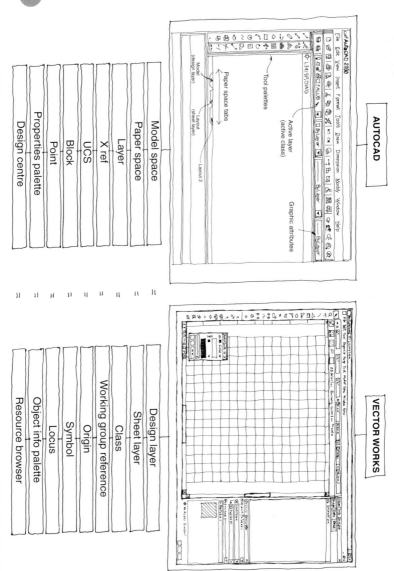

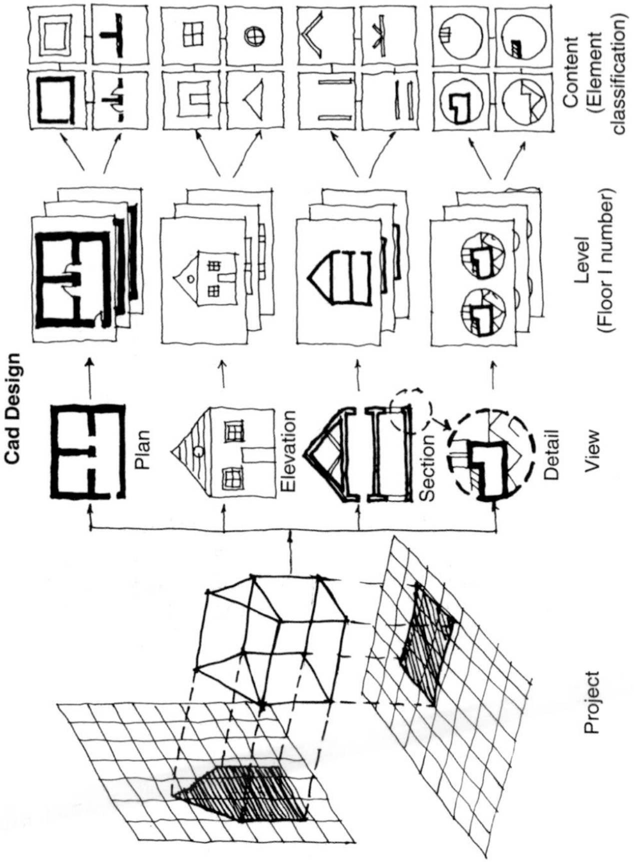
Cad Design
Plan
Elevation
Section
Detail
View
Level
(Floor I number)
Content
(Element
classification)
Project

Drawing production from cad data

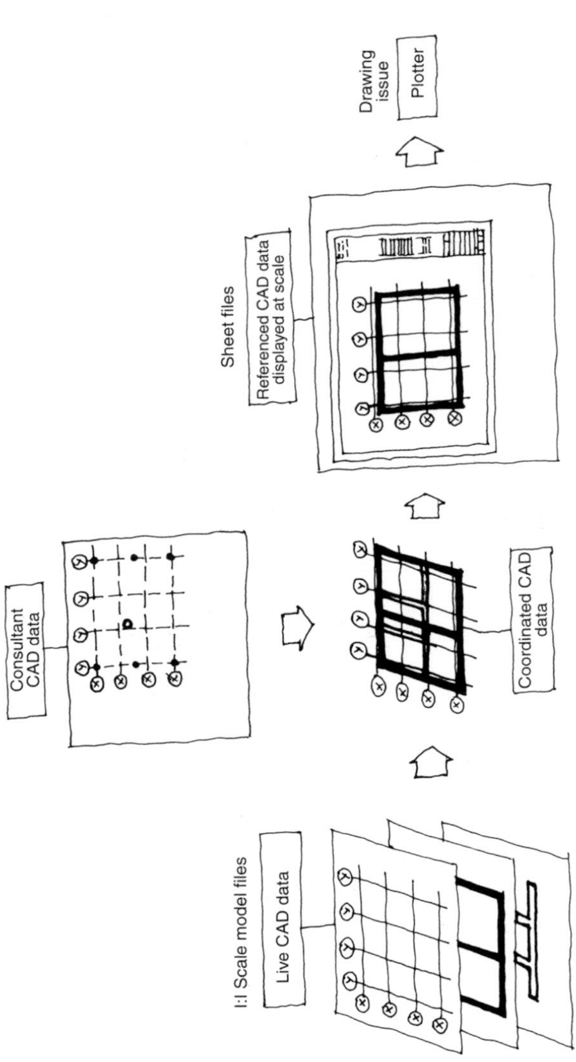
Consultant CAD data
I:I Scale model files
Live CAD data
Coordinated CAD data
Sheet files
Referenced CAD data displayed at scale
Drawing issue
Plotter

Classical temple

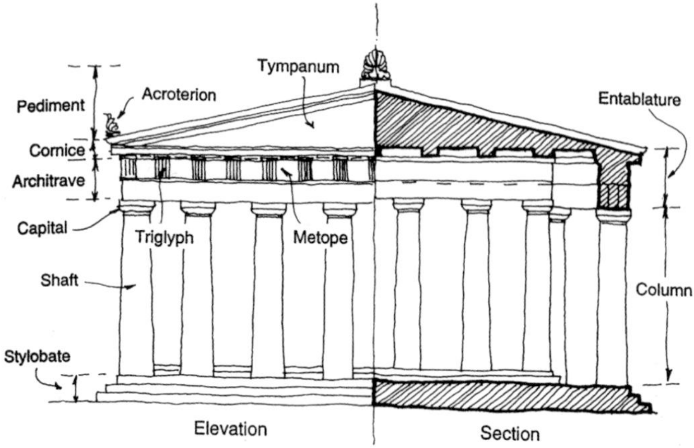
Tympanum
Pediment
Acroterion
Entablature
Cornice
Architrave
Capital
Triglyph
Metope
Shaft
Column
Stylobate
Elevation
Section

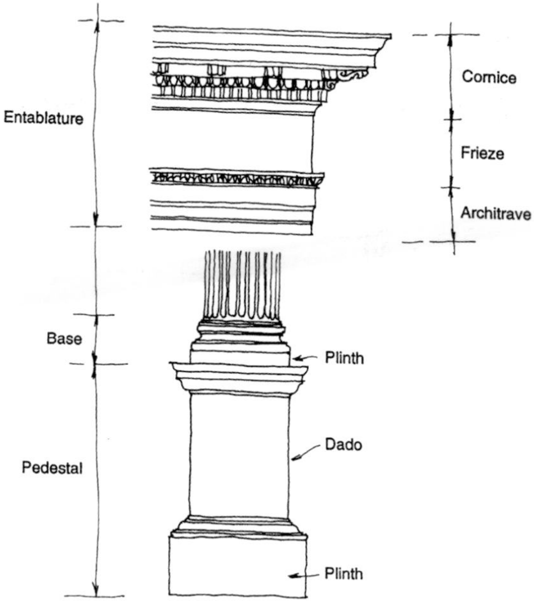
Entablature
Cornice
Frieze
Architrave
Base
Plinth
Dado
Pedestal
Plinth

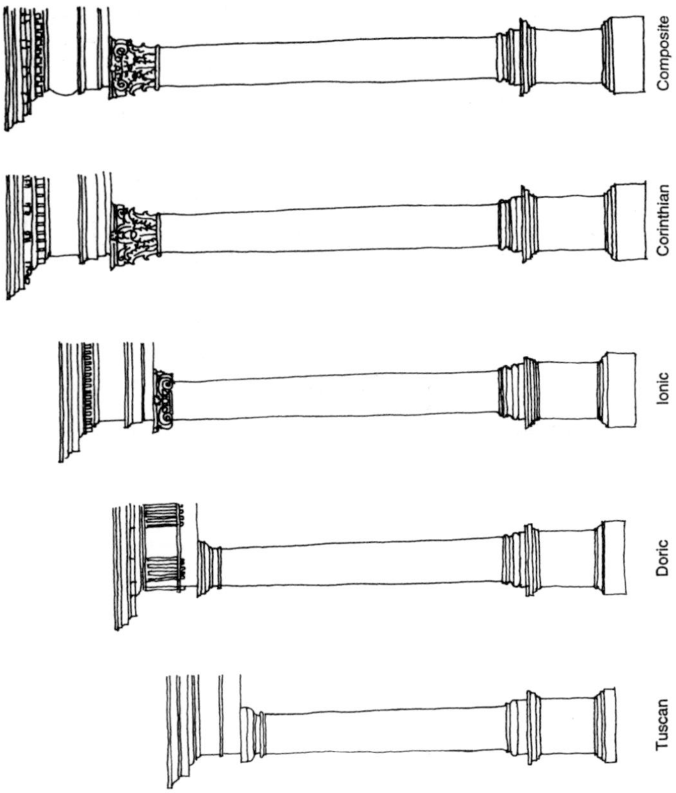
Composite
Corinthian
Ionic
Doric
Tuscan

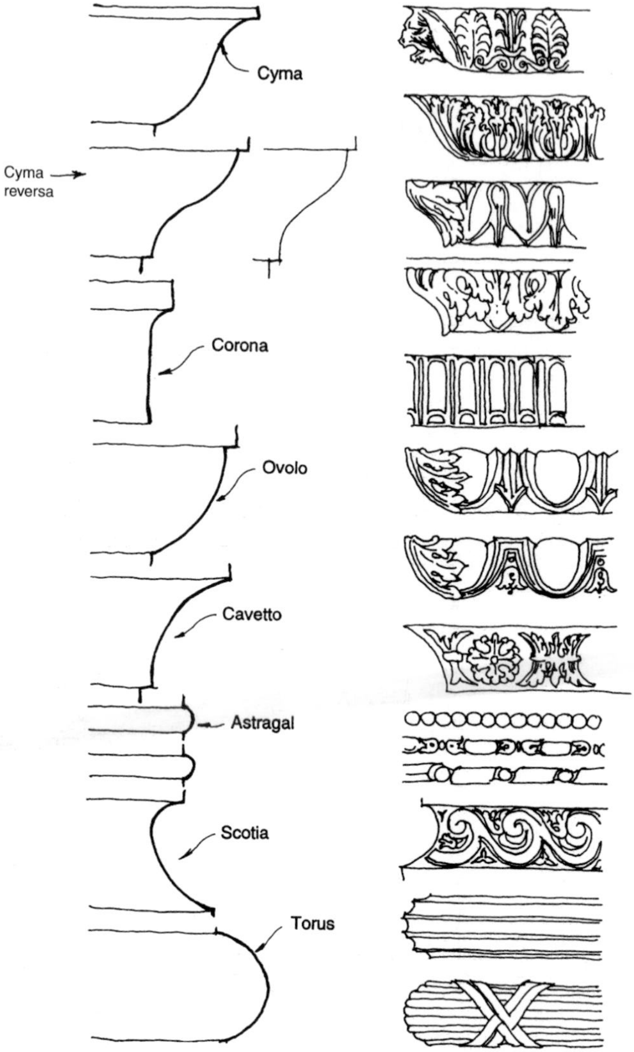

Cyma
Cyma reversa
Corona
Ovolo
Cavetto
Astragal
Scotia
Torus

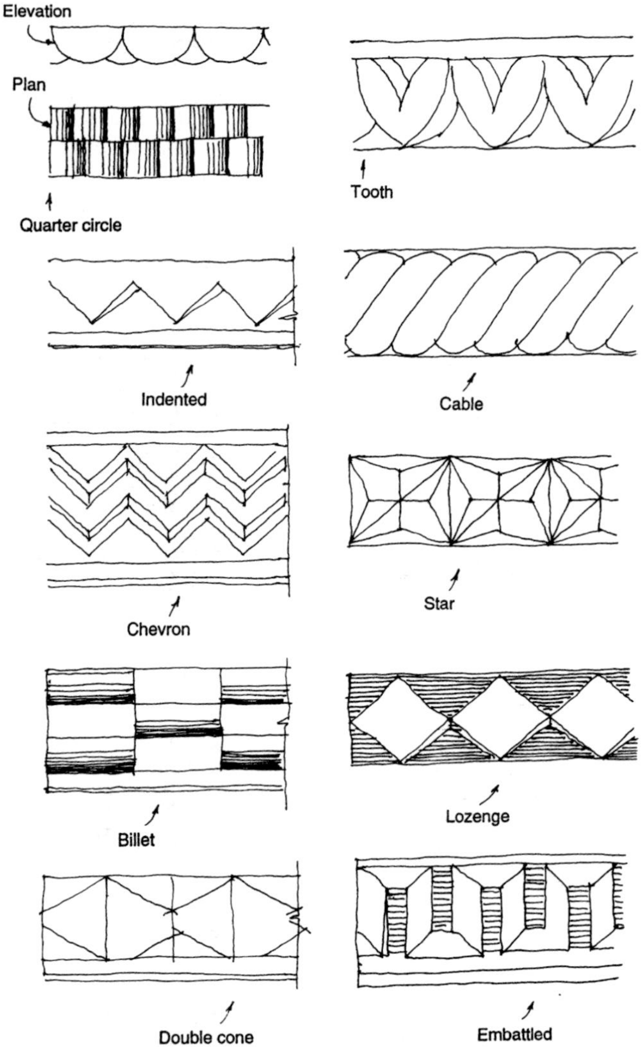
Elevation
Plan
Quarter circle
Tooth
Indented
Cable
Chevron
Star
Billet
Lozenge
Double cone
Embattled

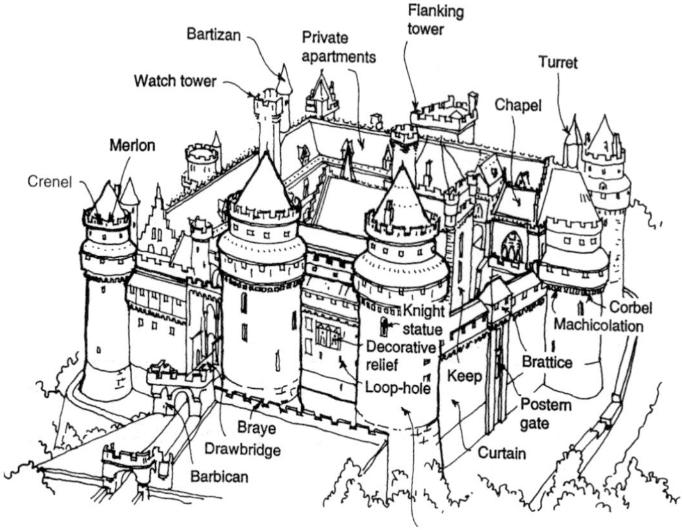
Flanking
tower
Bartizan
Private
apartments
Turret
Watch tower
Chapel
Merlon
Crenel
Knight
statue
Corbel
Machicolation
Decorative
relief
Brattice
Keep
Loop-hole
Postem
gate
Braye
Drawbridge
Curtain
Barbican
Corner tower

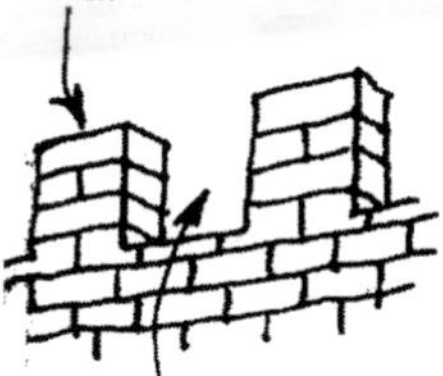
Merion
Crenel

Machicolation

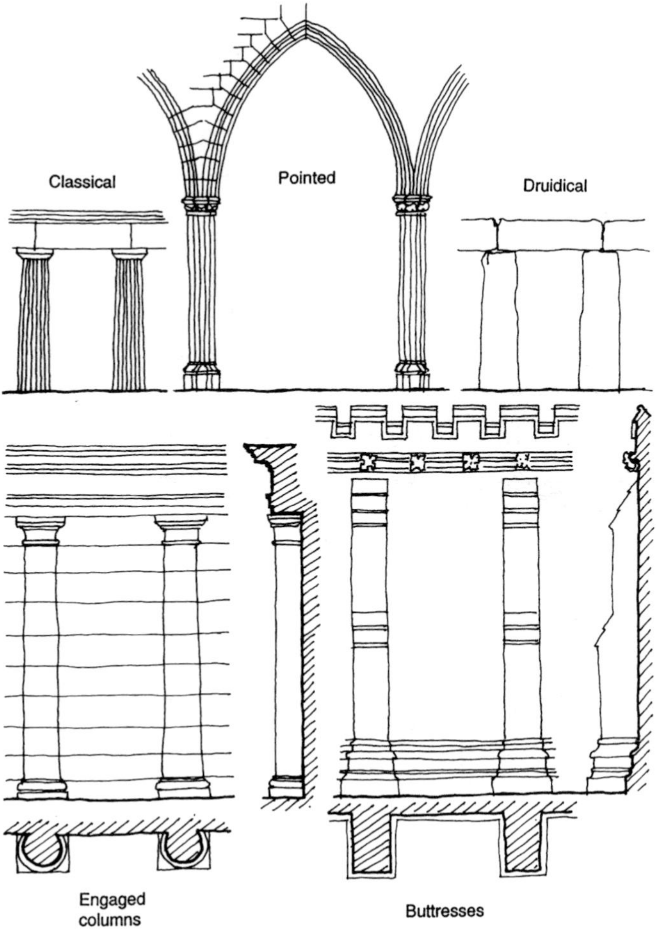
Classical
Pointed
Druidical
Engaged columns
Buttresses

Cruck framed bam

Timber framed
hall house

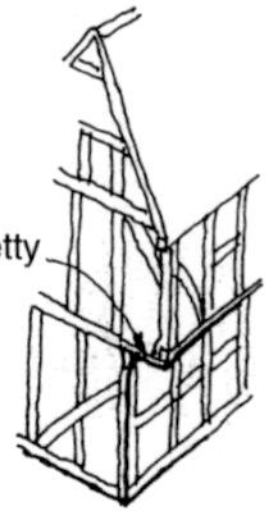
Jetty

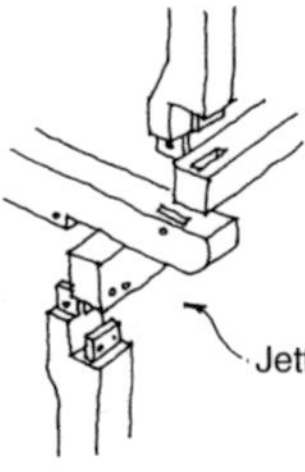
Jetty detail

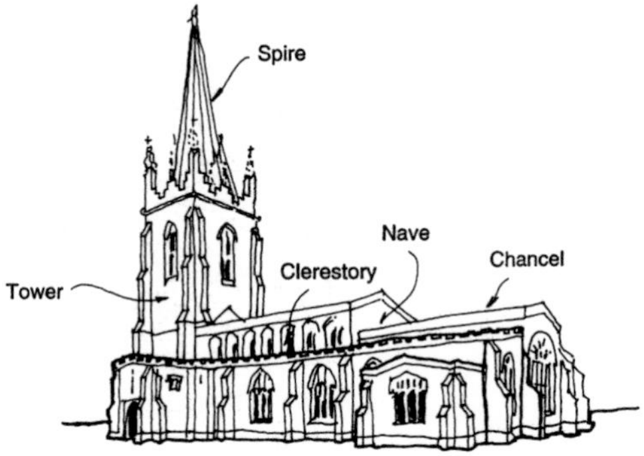
Spire
Tower
Nave
Clerestory
Chancel

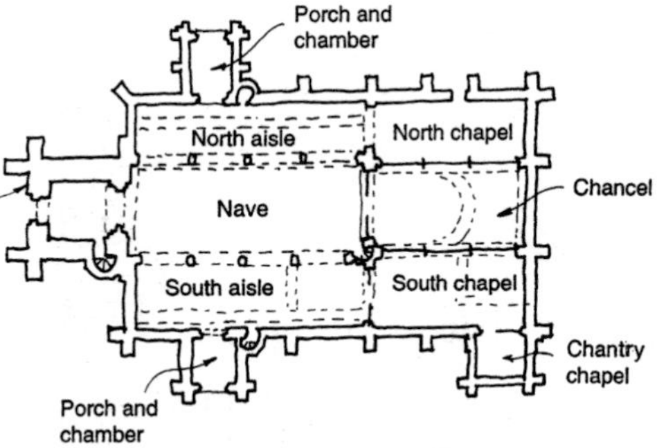
Porch and chamber
North aisle
North chapel
Chancel
Tower
Nave
South aisle
South chapel
Chantry chapel
Porch and chamber

Three-cell type plan

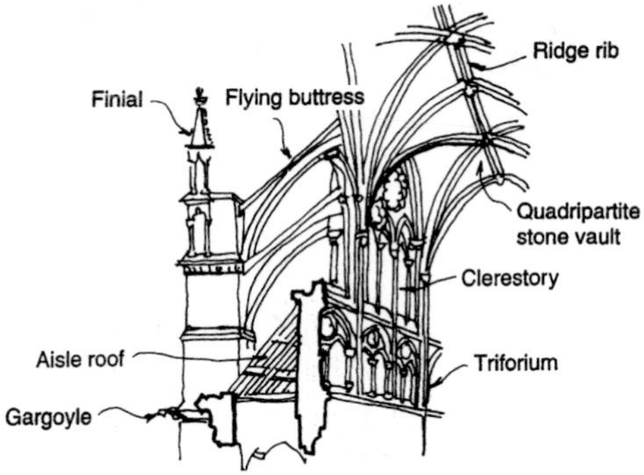
Ridge rib
Finial
Flying buttress
Quadripartite stone vault
Clerestory
Aisle roof
Triforium
Gargoyle

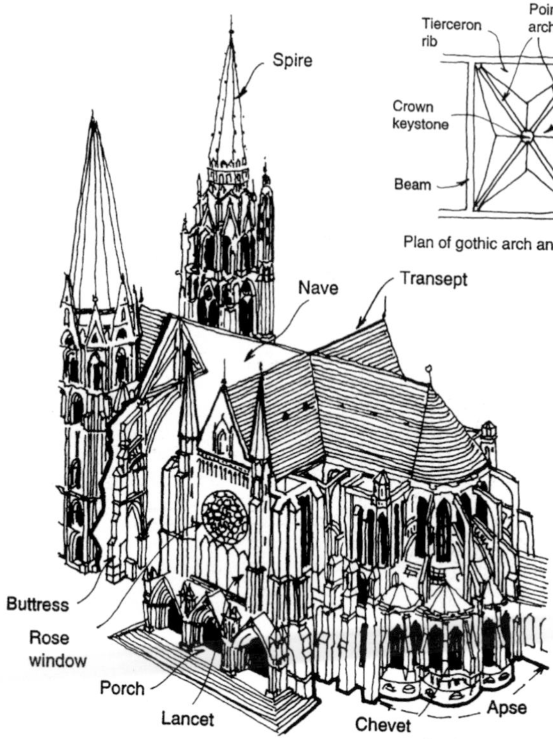
Spire
Nave
Transept
Buttress
Rose window
Porch
Lancet
Chevet
Apse

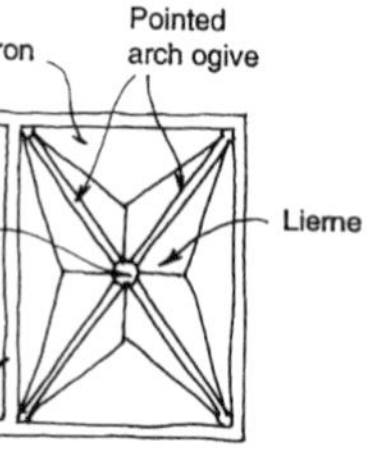
Tierceron rib
Pointed arch ogive
Crown keystone
Lierne
Beam

Plan of gothic arch and ribbed vault

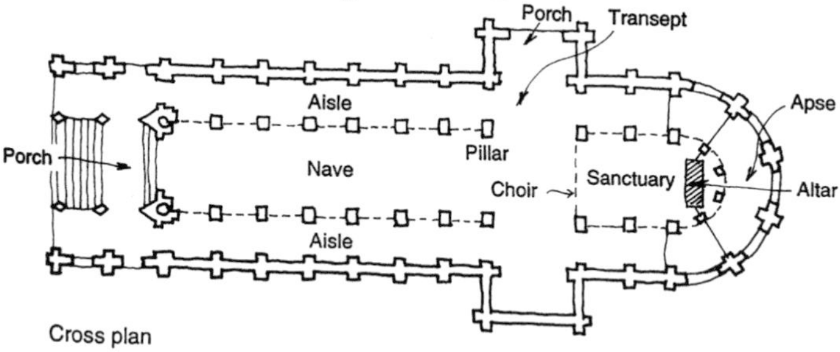
Porch
Transept
Apse
Aisle
Pillar
Porch
Nave
Choir
Sanctuary
Altar
Aisle

Cross plan

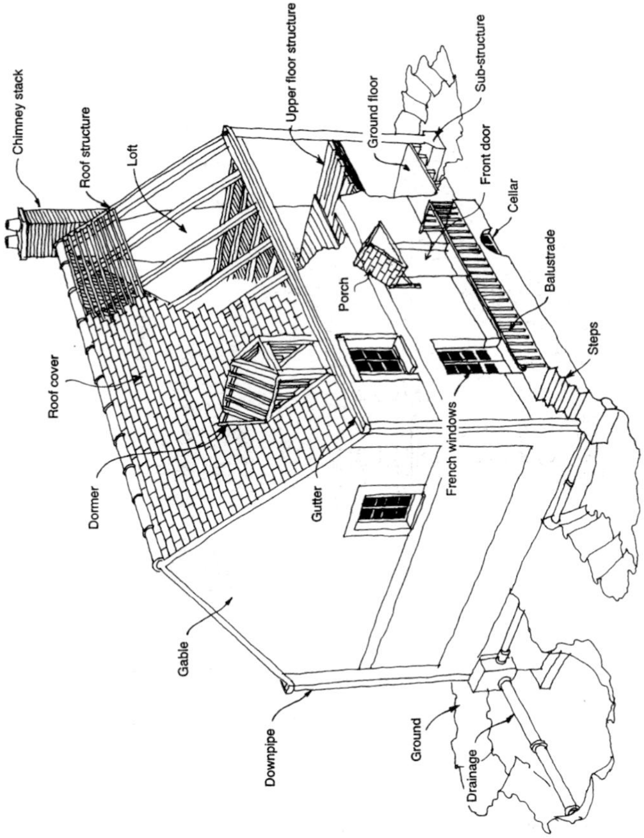

Chimney stack
Roof structure
Loft
Upper floor structure
Ground floor
Sub-structure
Front door
Cellar
Balustrade
Steps
Porch
Roof cover
French windows
Dormer
Gutter
Gable
Downpipe
Ground
Drainage

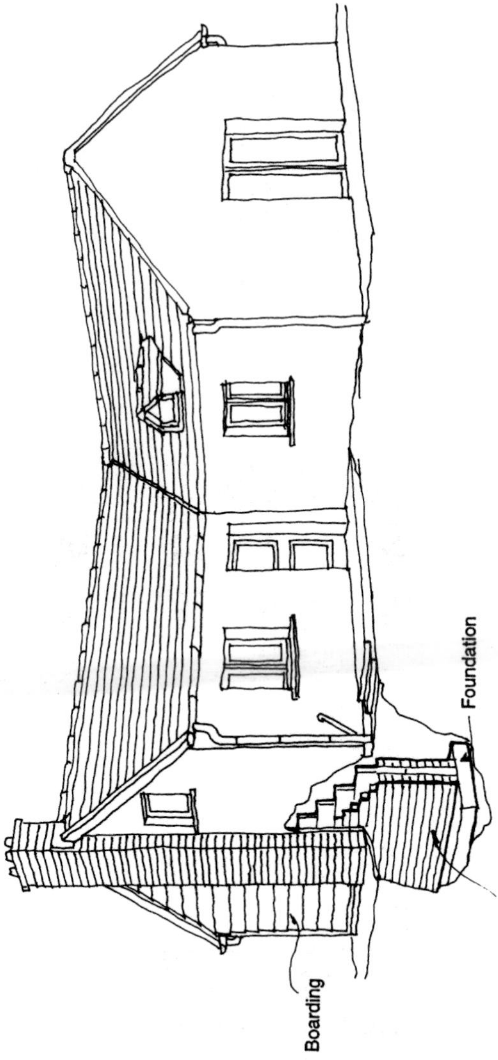
Boarding
Cavity wall
Foundation

Terraces

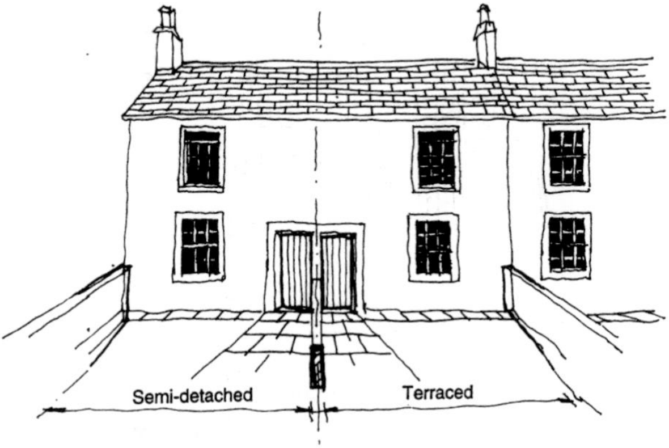

Detached cottage

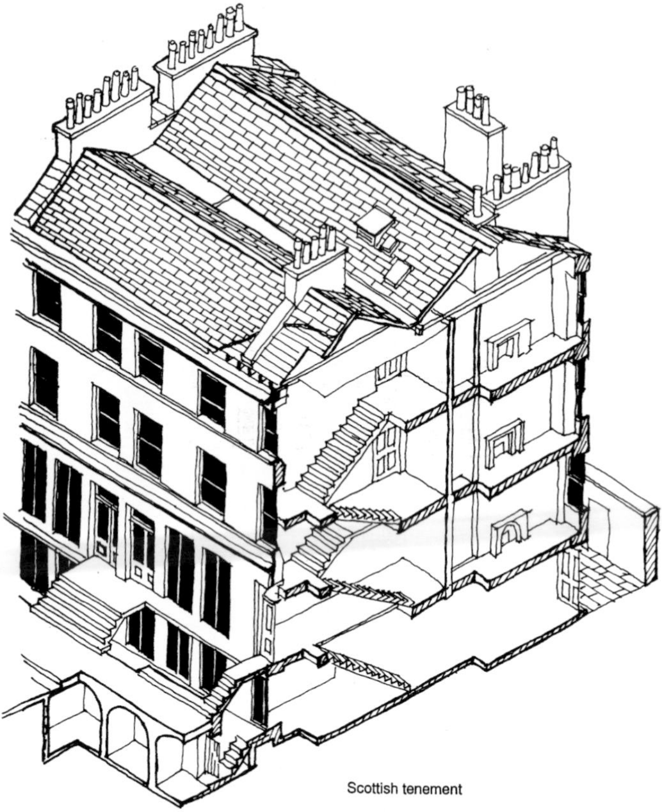

Scottish tenement

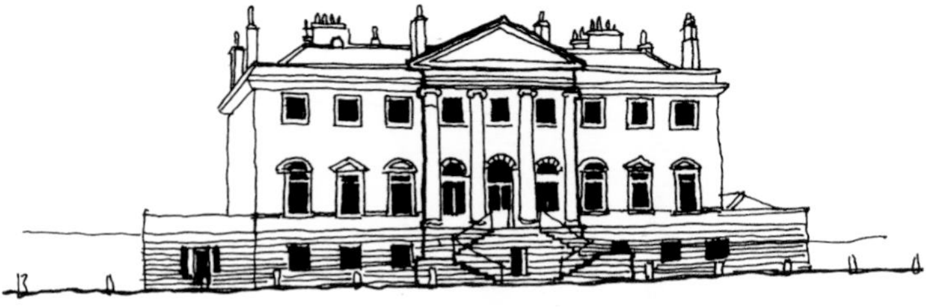

Stately home

Renaissance château

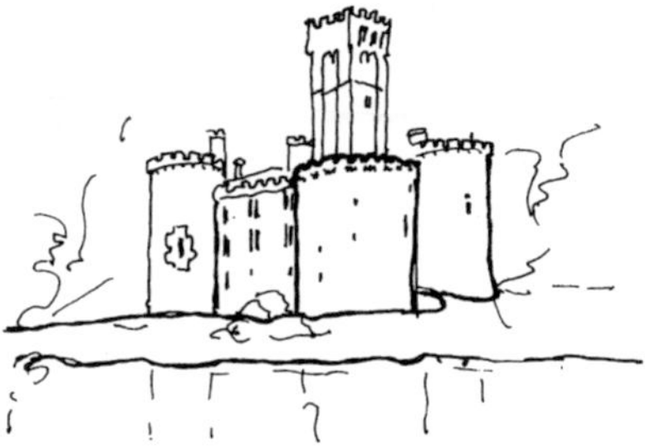

Mediaeval castle

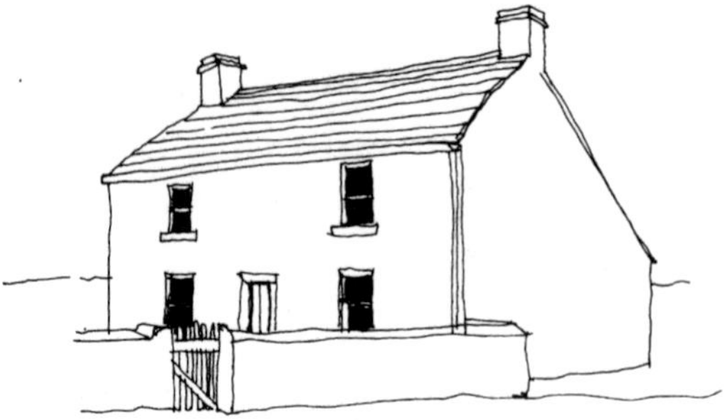

Detached house

Bungalow

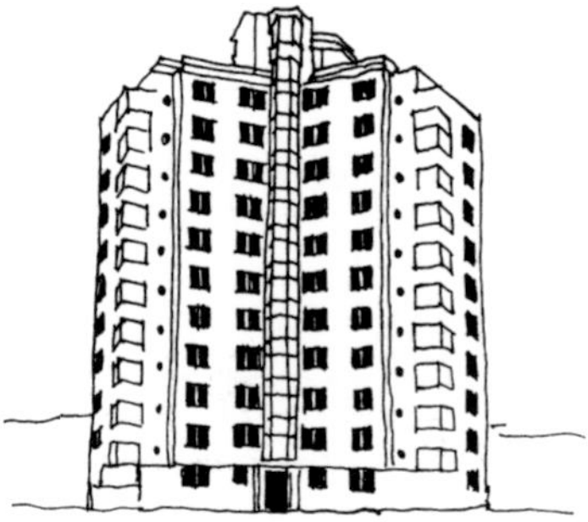

High-rise block of flats

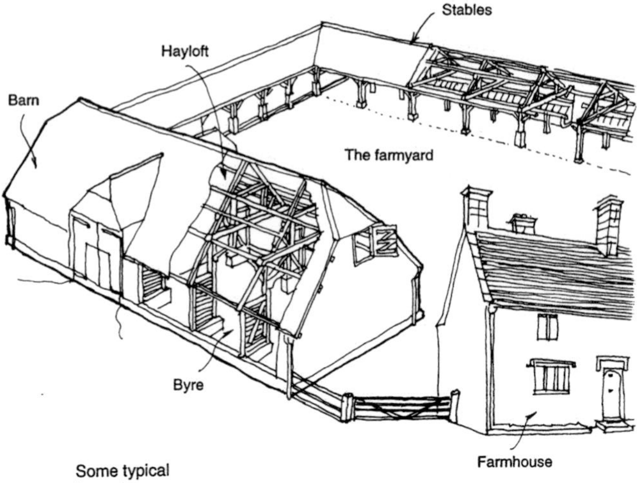
Stables
Hayloft
Barn
The farmyard
Byre
Farmhouse

Some typical barn shapes

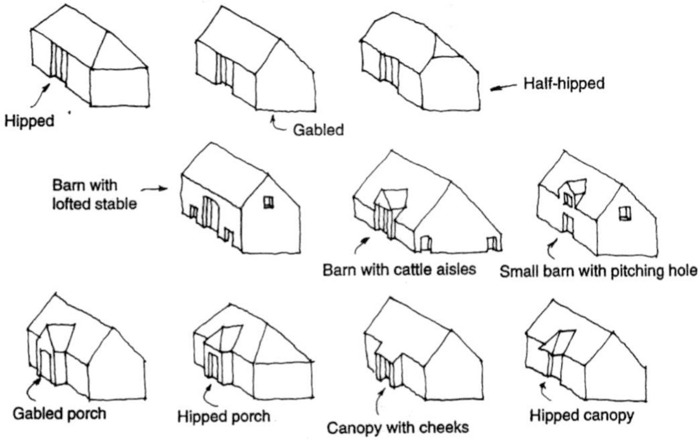
Hipped
Gabled
Half-hipped
Barn with lofted stable
Barn with cattle aisles
Small barn with pitching hole
Gabled porch
Hipped porch
Canopy with cheeks
Hipped canopy

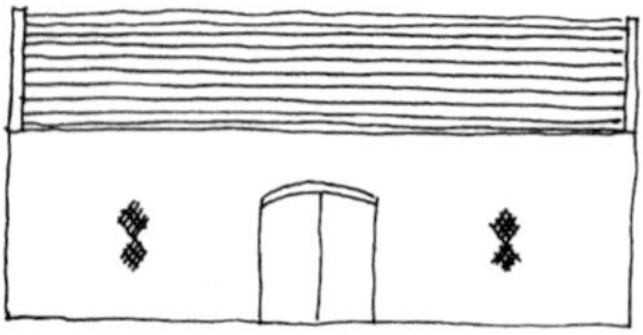
Three-bay
threshing beam

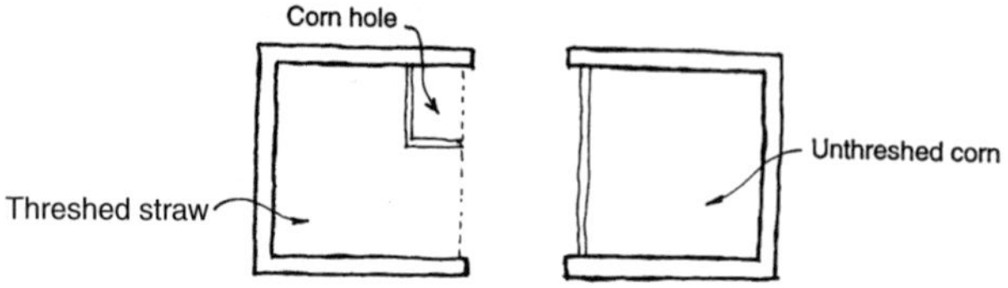
Corn hole
Threshed straw
Unthreshed corn

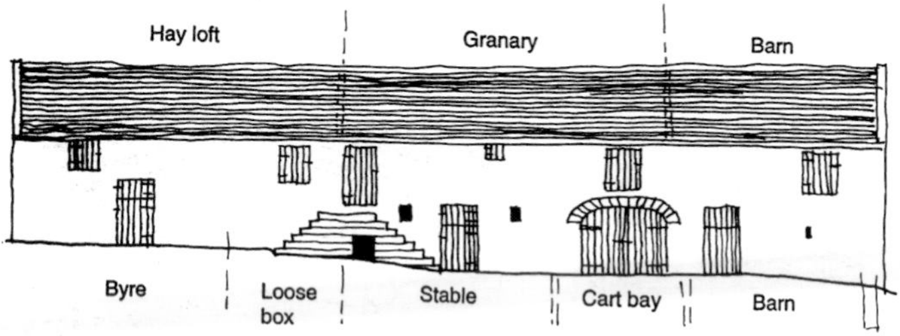
Hay loft
Granary
Barn
Byre
Loose box
Stable
Cart bay
Barn

Field house

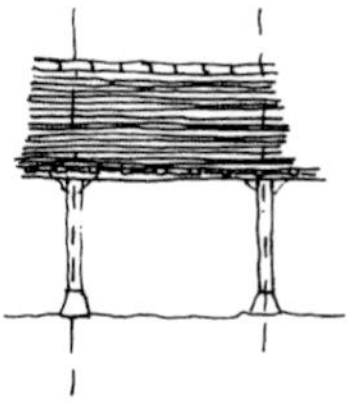
Open-fronted
shelter shed bay

Doorway

Doorcase

Door hood

Portal

Porch

Gate house

Gateway

Ambulatory

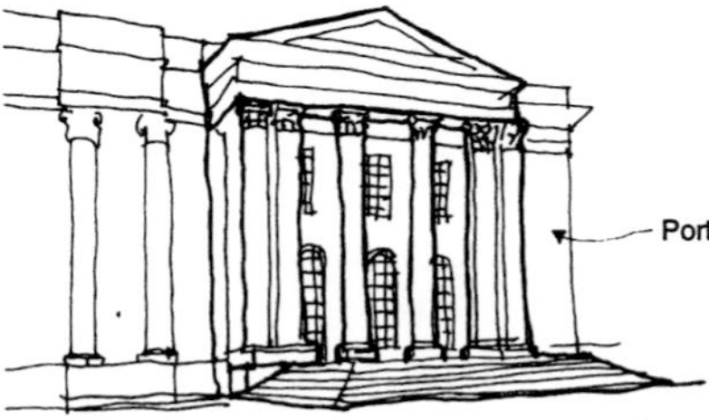
Portico

Porte-cochère

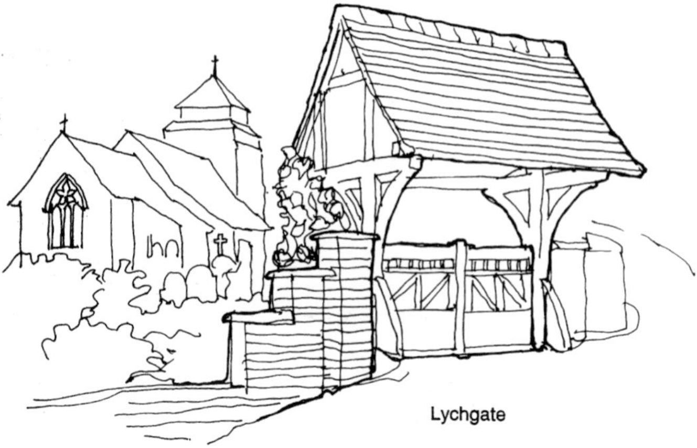

Lychgate

Loggia

Bay window

Bow window

Oriel window

Controls
Legal aspects
Administration

II. Controls

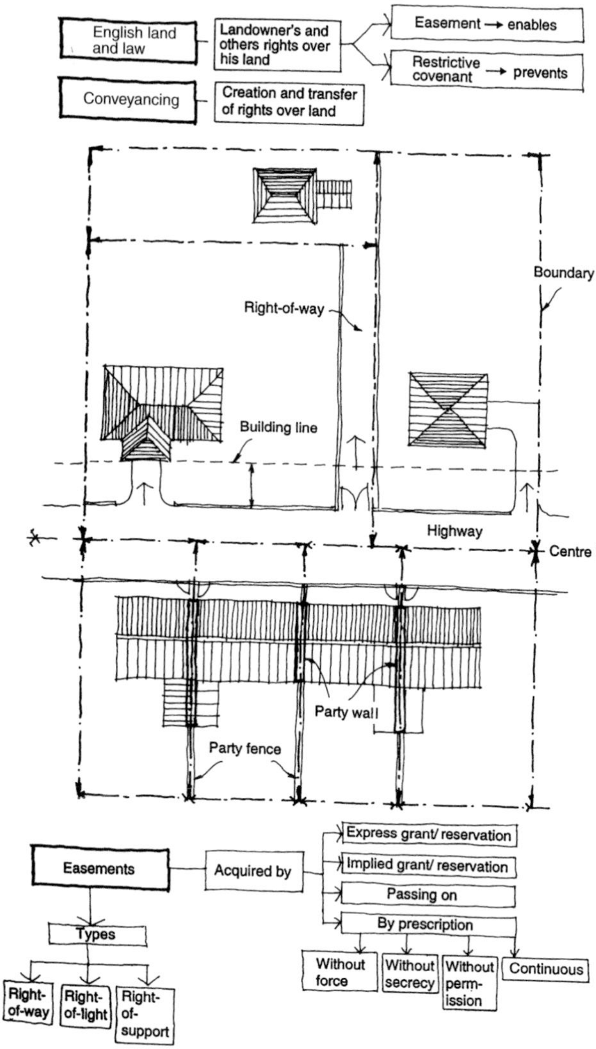
English land and law
Landowner's and others rights over his land
Easement → enables
Restrictive covenant → prevents
Conveyancing
Creation and transfer of rights over land
Right-of-way
Boundary line
Building line
Highway
Centre line
Party wall
Party fence
Easements
Acquired by
Express grant/ reservation
Implied grant/ reservation
Passing on
By prescription
Without force
Without secrecy
Without perm-ission
Continuous
Types
Right-of-way
Right-of-light
Right-of-support

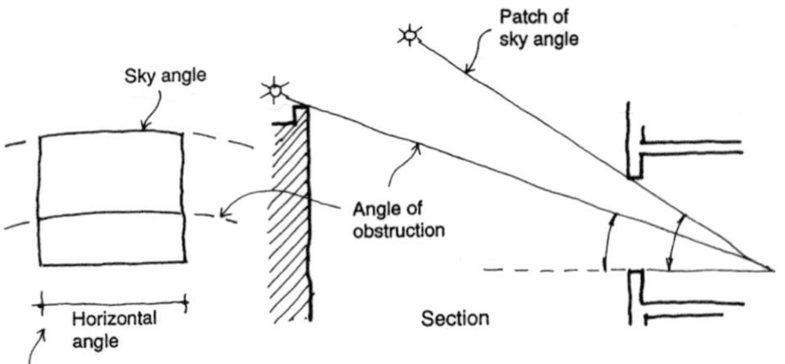

Sky factor (daylight from sky) defined by horizontal and vertical angles

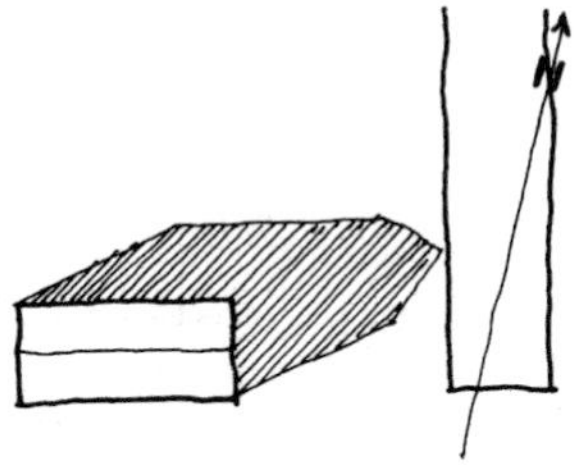

Sunlight indicator (shadow of buildings measurement)

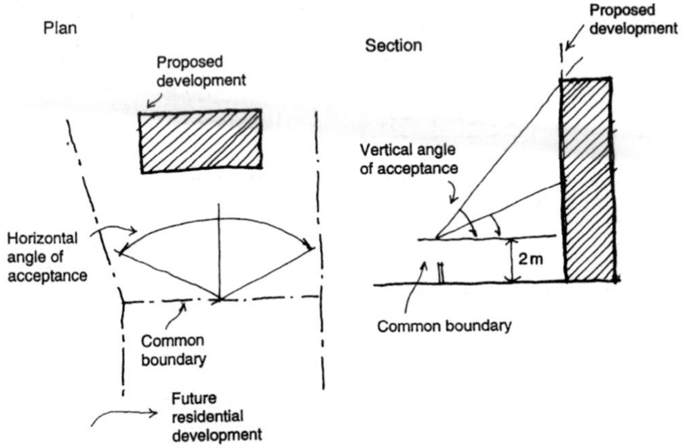

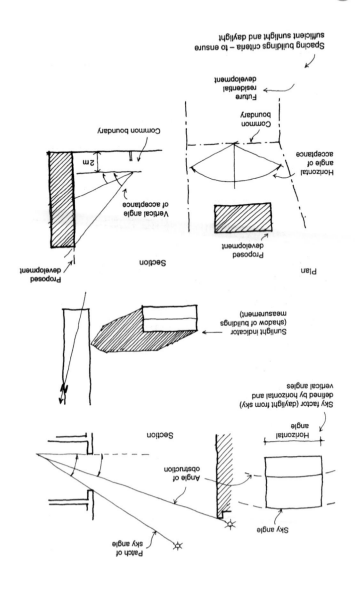

Spacing buildings criteria – to ensure sufficient sunlight and daylight

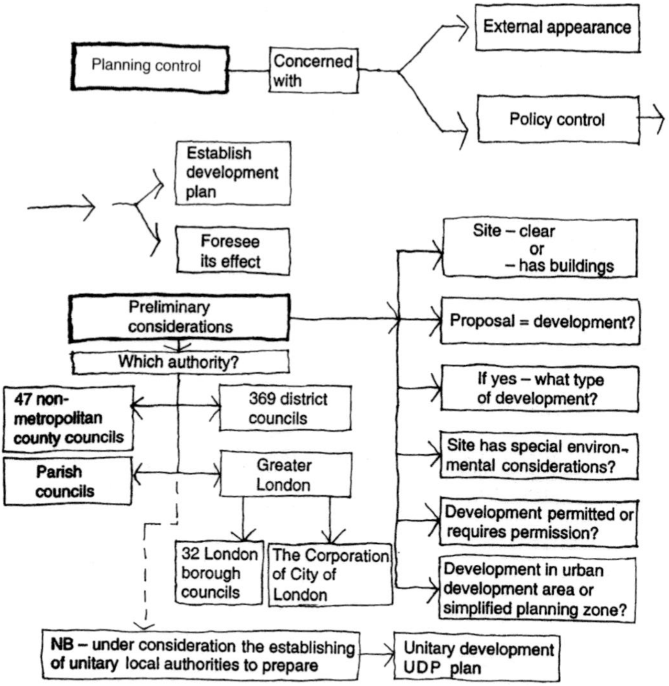
Planning control
Concerned with
External appearance
Policy control
Establish development plan
Foresee its effect
Preliminary considerations
Which authority?
47 non-metropolitan county councils
369 district councils
Parish councils
Greater London
32 London borough councils
The Corporation of City of London
NB – under consideration the establishing of unitary local authorities to prepare
Unitary development UDP plan
Site – clear or – has buildings
Proposal = development?
If yes – what type of development?
Site has special environmental considerations?
Development permitted or requires permission?
Development in urban development area or simplified planning zone?

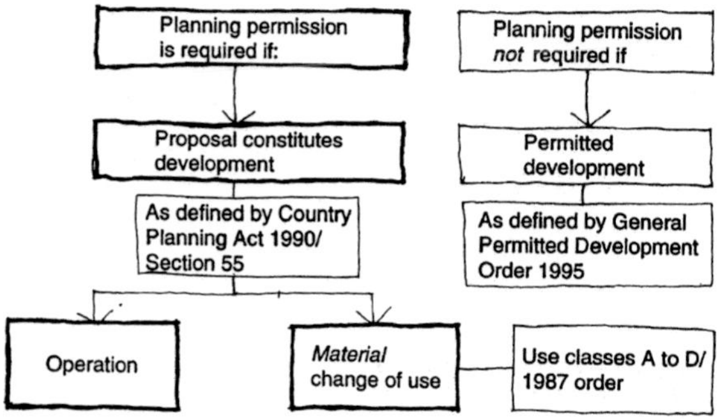
Planning permission is required if:
Proposal constitutes development
As defined by Country Planning Act 1990/ Section 55
Operation
Material change of use
Use classes A to D/ 1987 order
Planning permission not required if
Permitted development
As defined by General Permitted Development Order 1995

Planning approvals

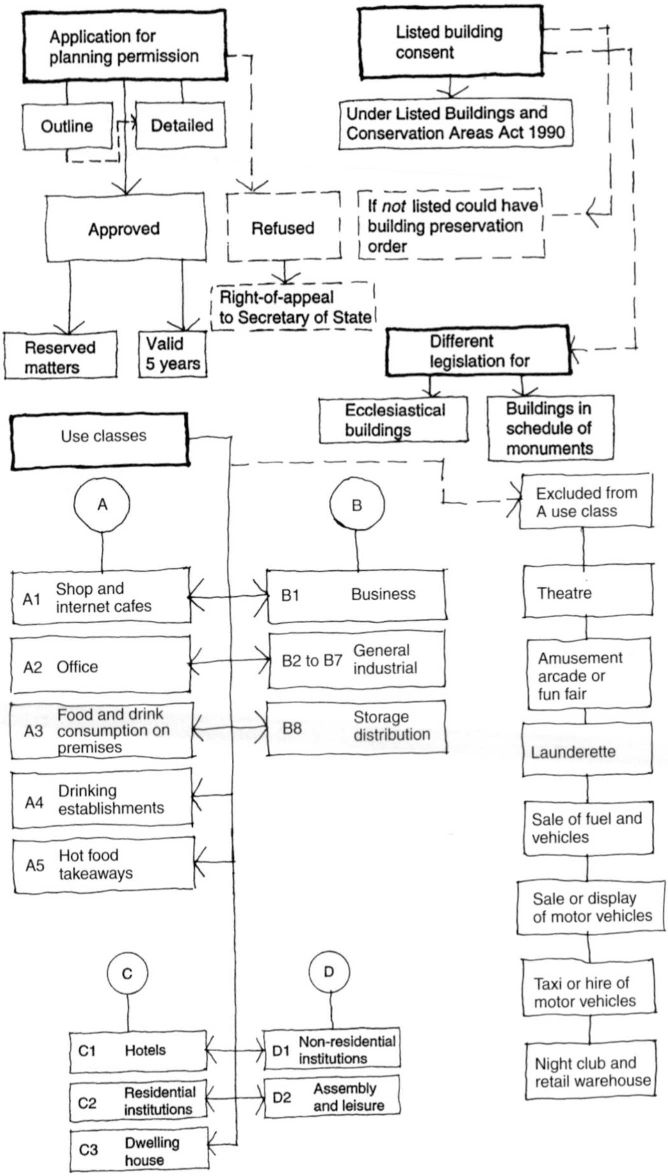

Application for planning permission
Outline
Detailed
Approved
Refused
Right-of-appeal to Secretary of State
Reserved matters
Valid 5 years
Listed building consent
Under Listed Buildings and Conservation Areas Act 1990
If *not* listed could have building preservation order
Different legislation for
Ecclesiastical buildings
Buildings in schedule of monuments
Use classes
A
B
A1 Shop and internet cafes
A2 Office
A3 Food and drink consumption on premises
A4 Drinking establishments
A5 Hot food takeaways
B1 Business
B2 to B7 General industrial
B8 Storage distribution
Excluded from A use class
Theatre
Amusement arcade or fun fair
Launderette
Sale of fuel and vehicles
Sale or display of motor vehicles
Taxi or hire of motor vehicles
Night club and retail warehouse
C
D
C1 Hotels
C2 Residential institutions
C3 Dwelling house
D1 Non-residential institutions
D2 Assembly and leisure

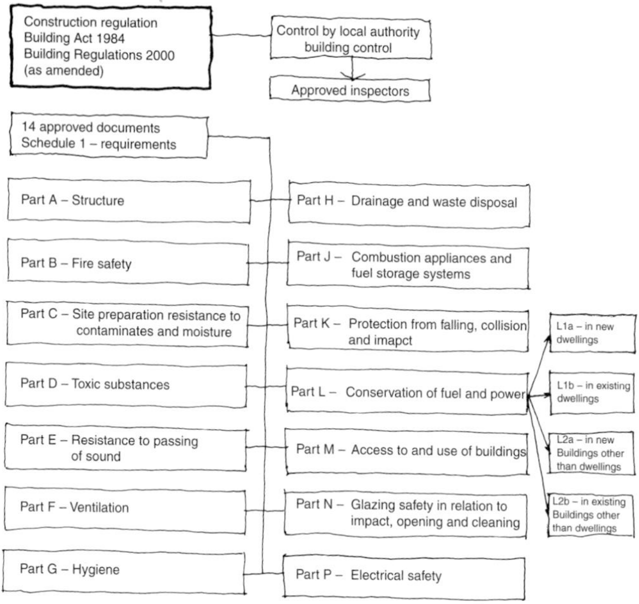

Subject	Officer	Local authority
Planning	Planning Officer	DC
Building regs.	Building Inspector	DC
Development in private street	Surveyor	CC
Surface water	Engineer	SU
Sewer connections	Engineer	SU
Blocked sewer	Engineer	SU
Housing, grants	Env. Health Officer	DC
Licensing	Petroleum Insp. + Env. H.	CC/DC
Height of chimneys	Env. Health Officer	DC
Music & dance licence	Local magistrate clerk	
Liquor licence	Local magistrate clerk	

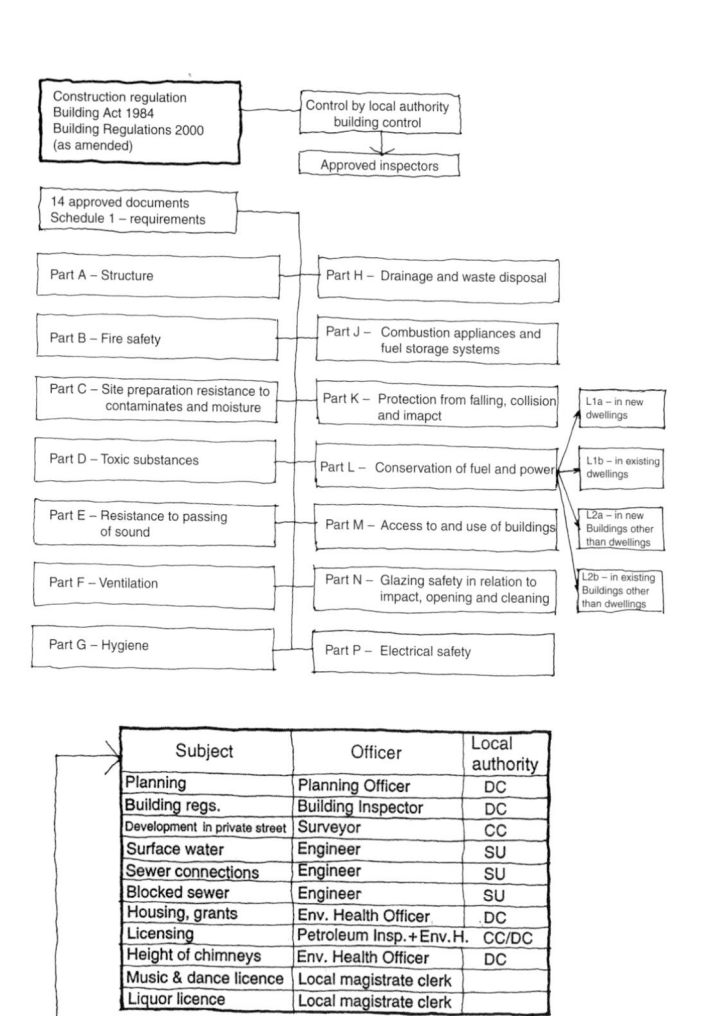

Responsibilities of local authority offices

Building control in London

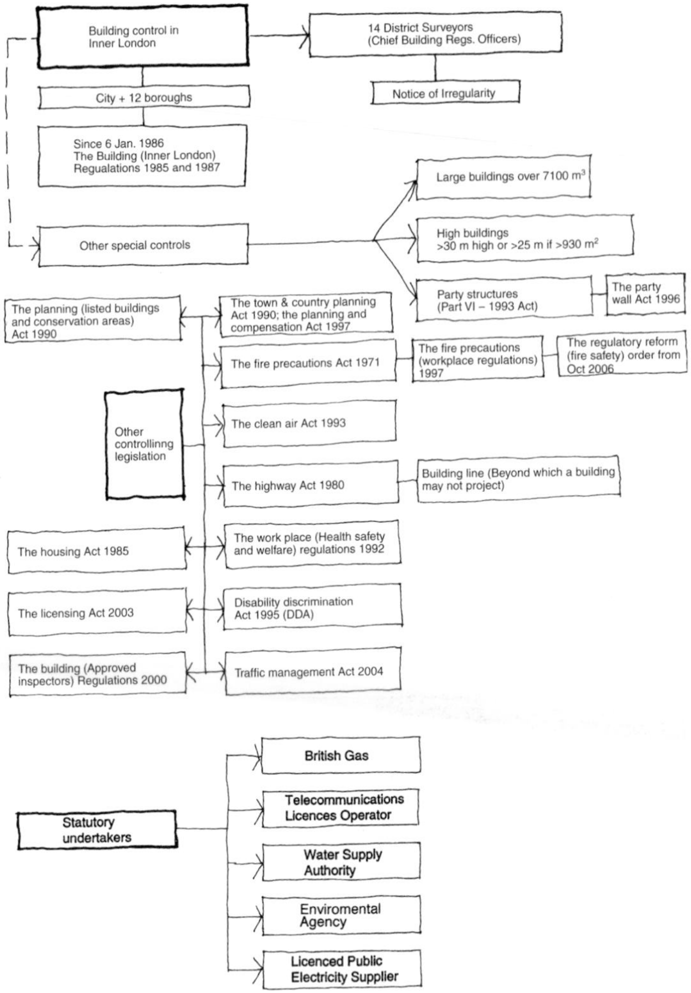
Building control in Inner London
14 District Surveyors (Chief Building Regs. Officers)
Notice of Irregularity
City + 12 boroughs
Since 6 Jan. 1986 The Building (Inner London) Regualations 1985 and 1987
Other special controls
Large buildings over 7100 m³
High buildings >30 m high or >25 m if >930 m²
Party structures (Part VI – 1993 Act)
The party wall Act 1996
Other controllinng legislation
The planning (listed buildings and conservation areas) Act 1990
The town & country planning Act 1990; the planning and compensation Act 1997
The fire precautions Act 1971
The fire precautions (workplace regulations) 1997
The regulatory reform (fire safety) order from Oct 2006
The clean air Act 1993
The highway Act 1980
Building line (Beyond which a building may not project)
The housing Act 1985
The work place (Health safety and welfare) regulations 1992
The licensing Act 2003
Disability discrimination Act 1995 (DDA)
The building (Approved inspectors) Regulations 2000
Traffic management Act 2004
Statutory undertakers
British Gas
Telecommunications Licences Operator
Water Supply Authority
Enviromental Agency
Licenced Public Electricity Supplier

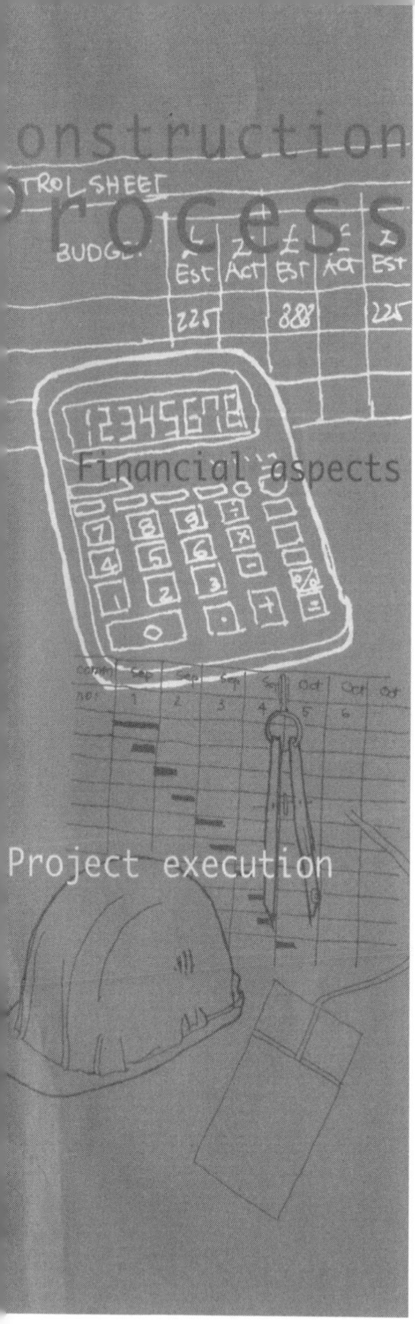

III. Construction Process

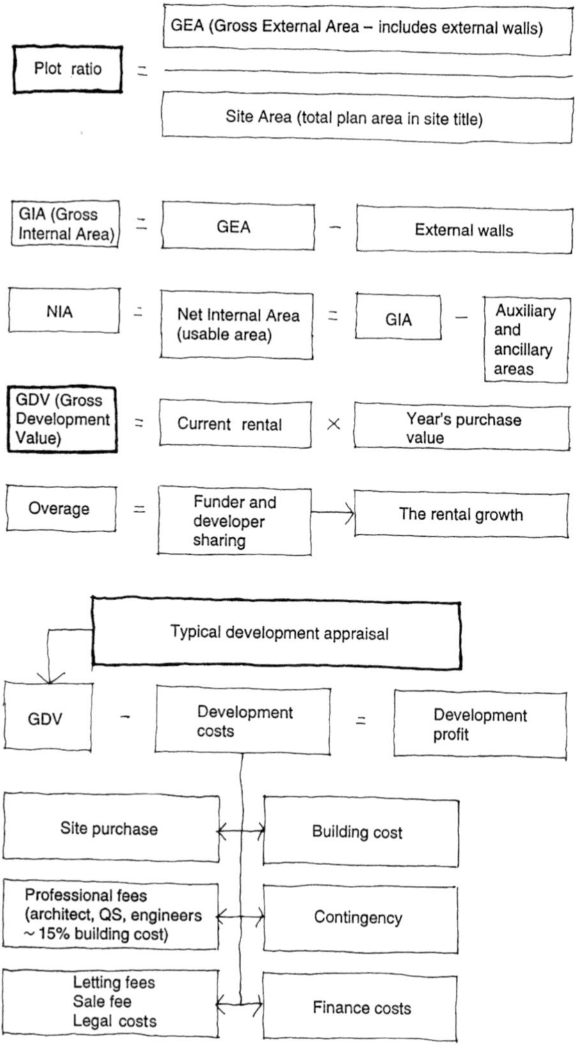
Plot ratio
=
GEA (Gross External Area – includes external walls)
Site Area (total plan area in site title)
GIA (Gross Internal Area)
=
GEA
–
External walls
NIA
=
Net Internal Area (usable area)
=
GIA
–
Auxiliary and ancillary areas
GDV (Gross Development Value)
=
Current rental
×
Year's purchase value
Overage
=
Funder and developer sharing
The rental growth
Typical development appraisal
GDV
–
Development costs
=
Development profit
Site purchase
Building cost
Professional fees (architect, QS, engineers ~ 15% building cost)
Contingency
Letting fees
Sale fee
Legal costs
Finance costs

Development funding methods

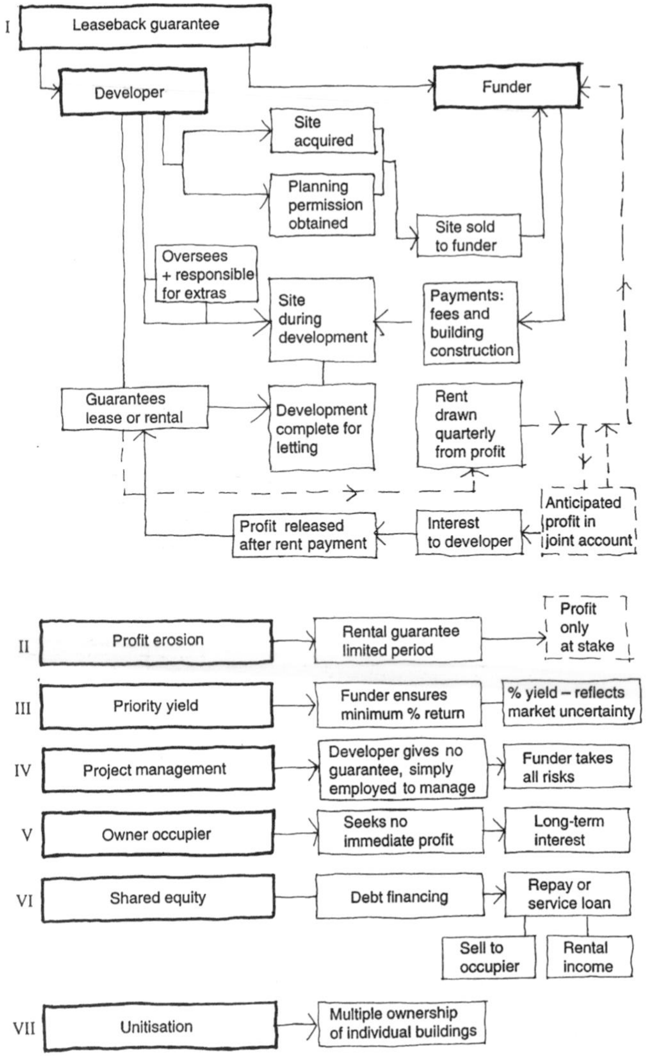
I
Leaseback guarantee
Developer
Funder
Site acquired
Planning permission obtained
Site sold to funder
Oversees + responsible for extras
Site during development
Payments: fees and building construction
Guarantees lease or rental
Development complete for letting
Rent drawn quarterly from profit
Anticipated profit in joint account
Profit released after rent payment
Interest to developer
II
Profit erosion
Rental guarantee limited period
Profit only at stake
III
Priority yield
Funder ensures minimum % return
% yield – reflects market uncertainty
IV
Project management
Developer gives no guarantee, simply employed to manage
Funder takes all risks
V
Owner occupier
Seeks no immediate profit
Long-term interest
VI
Shared equity
Debt financing
Repay or service loan
Sell to occupier
Rental income
VII
Unitisation
Multiple ownership of individual buildings

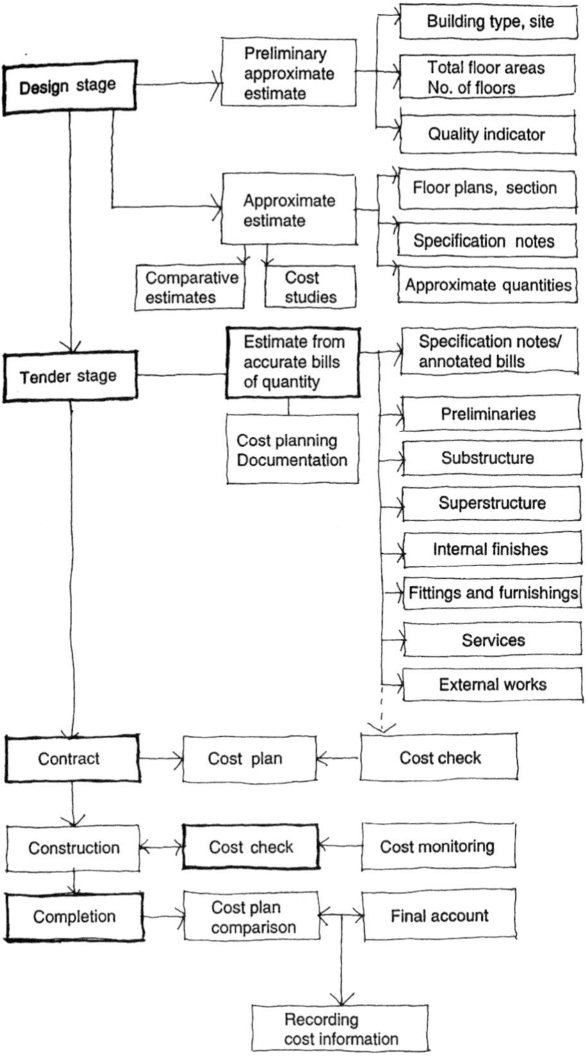

Design stage
Preliminary approximate estimate
Building type, site
Total floor areas
No. of floors
Quality indicator
Approximate estimate
Floor plans, section
Specification notes
Approximate quantities
Comparative estimates
Cost studies
Tender stage
Estimate from accurate bills of quantity
Cost planning Documentation
Specification notes/ annotated bills
Preliminaries
Substructure
Superstructure
Internal finishes
Fittings and furnishings
Services
External works
Contract
Cost plan
Cost check
Construction
Cost check
Cost monitoring
Completion
Cost plan comparison
Final account
Recording cost information

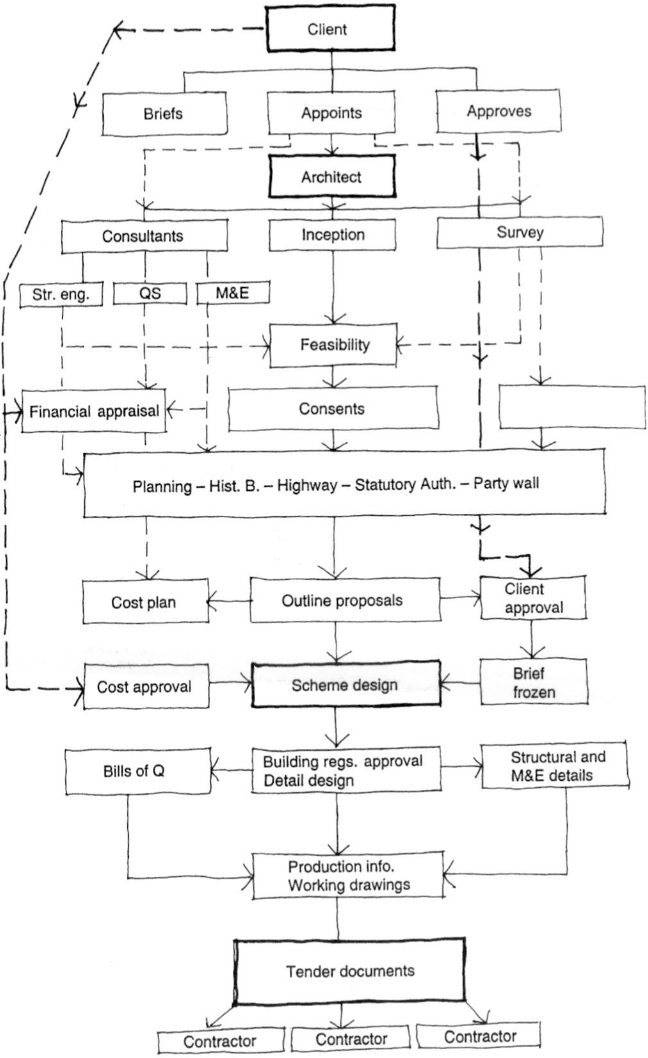
Client
Briefs
Appoints
Approves
Architect
Consultants
Inception
Survey
Str. eng.
QS
M&E
Feasibility
Financial appraisal
Consents
Planning – Hist. B. – Highway – Statutory Auth. – Party wall
Cost plan
Outline proposals
Client approval
Cost approval
Scheme design
Brief frozen
Bills of Q
Building regs. approval
Detail design
Structural and M&E details
Production info.
Working drawings
Tender documents
Contractor
Contractor
Contractor

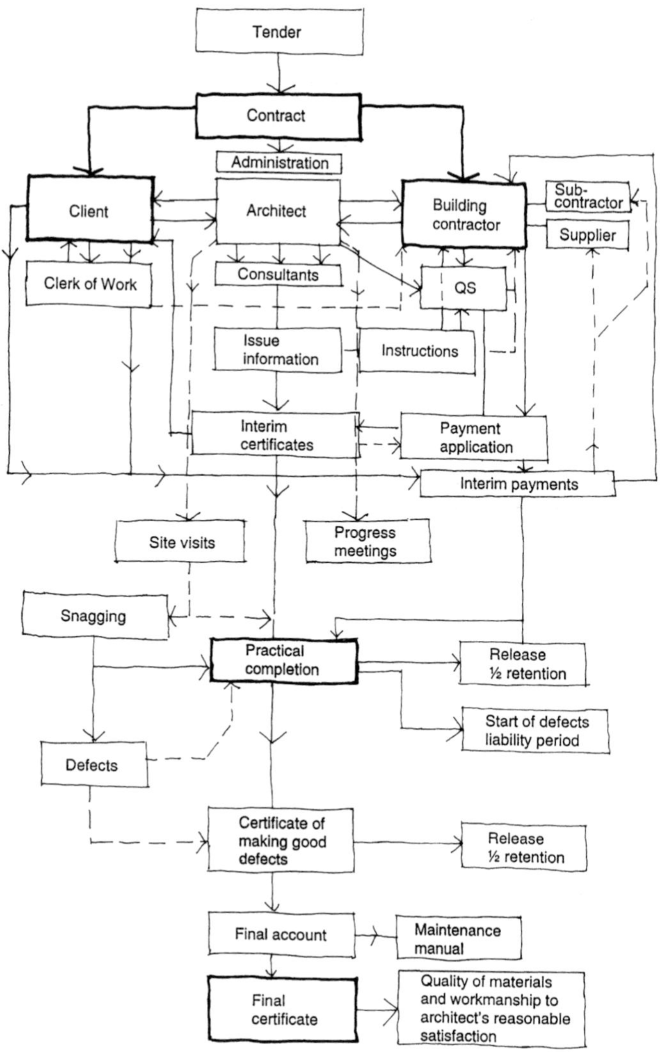
Tender
Contract
Administration
Client
Architect
Building contractor
Sub-contractor
Supplier
Clerk of Work
Consultants
QS
Issue information
Instructions
Interim certificates
Payment application
Interim payments
Site visits
Progress meetings
Snagging
Practical completion
Release ½ retention
Start of defects liability period
Defects
Certificate of making good defects
Release ½ retention
Final account
Maintenance manual
Final certificate
Quality of materials and workmanship to architect's reasonable satisfaction

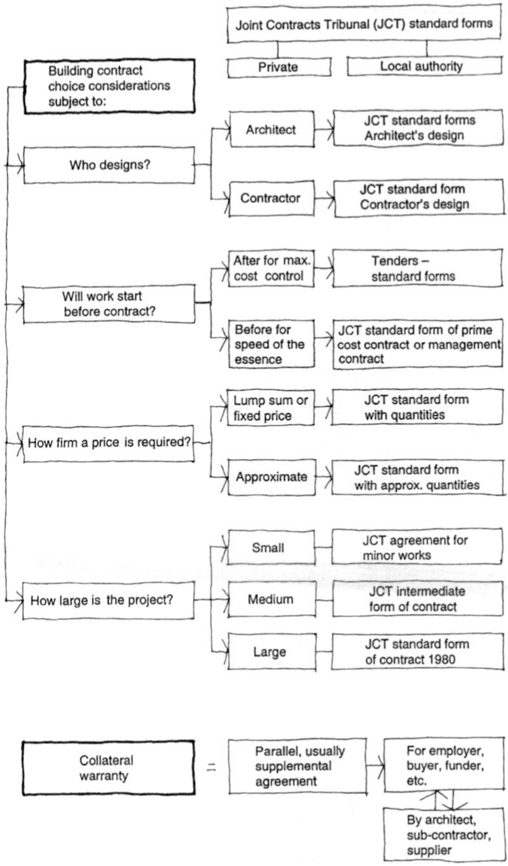
Joint Contracts Tribunal (JCT) standard forms
Private
Local authority
Building contract choice considerations subject to:
Who designs?
Architect
JCT standard forms Architect's design
Contractor
JCT standard form Contractor's design
Will work start before contract?
After for max. cost control
Tenders – standard forms
Before for speed of the essence
JCT standard form of prime cost contract or management contract
How firm a price is required?
Lump sum or fixed price
JCT standard form with quantities
Approximate
JCT standard form with approx. quantities
How large is the project?
Small
JCT agreement for minor works
Medium
JCT intermediate form of contract
Large
JCT standard form of contract 1980
Collateral warranty
Parallel, usually supplemental agreement
For employer, buyer, funder, etc.
By architect, sub-contractor, supplier

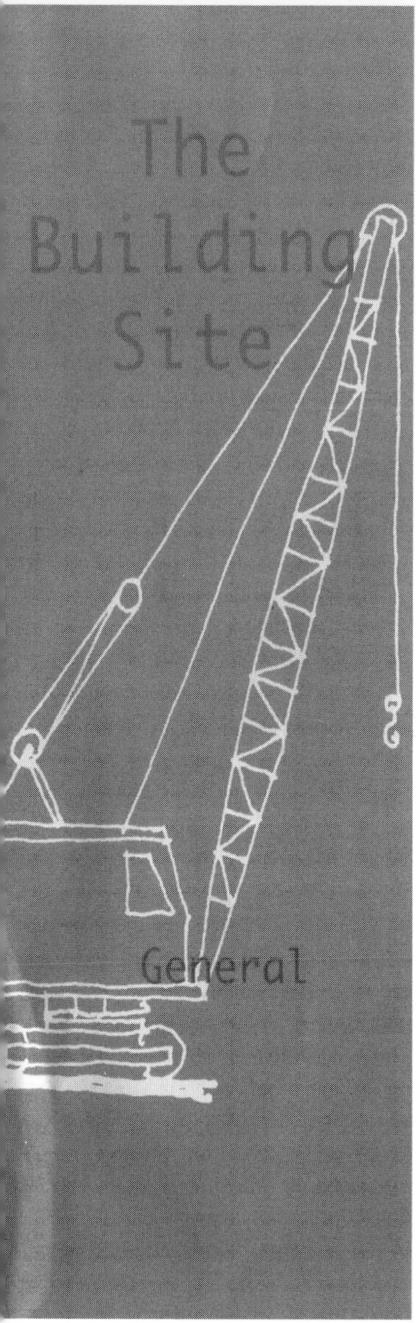

IV. The Building Site

General

Setting out building outline

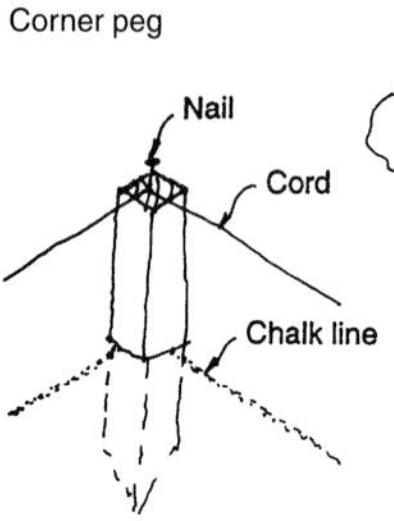
Corner peg
Nail
Cord
Chalk line

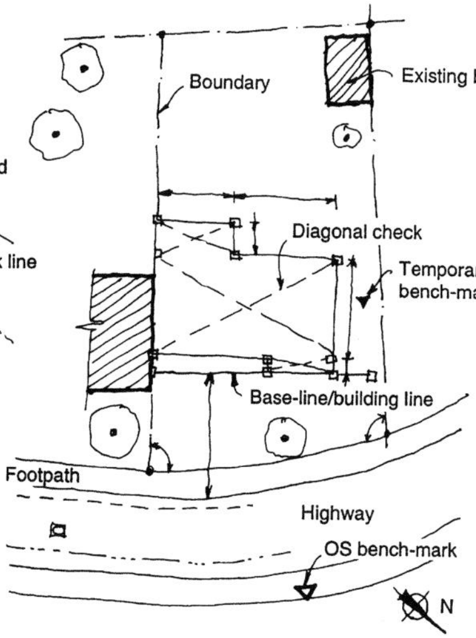
Boundary
Existing building
Diagonal check
Temporary bench-mark
Base-line/building line
Footpath
Highway
OS bench-mark
N

Setting out grid

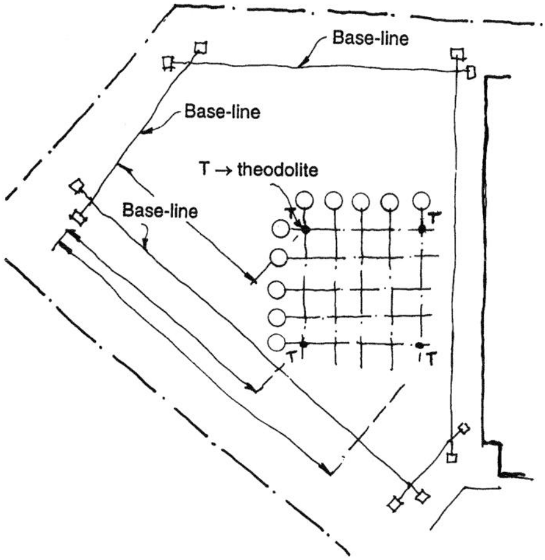
Base-line
Base-line
T → theodolite
Base-line

Setting out site levels

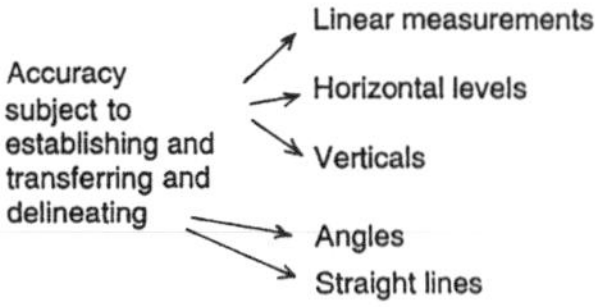
Accuracy
subject to
establishing and
transferring and
delineating
Linear measurements
Horizontal levels
Verticals
Angles
Straight lines

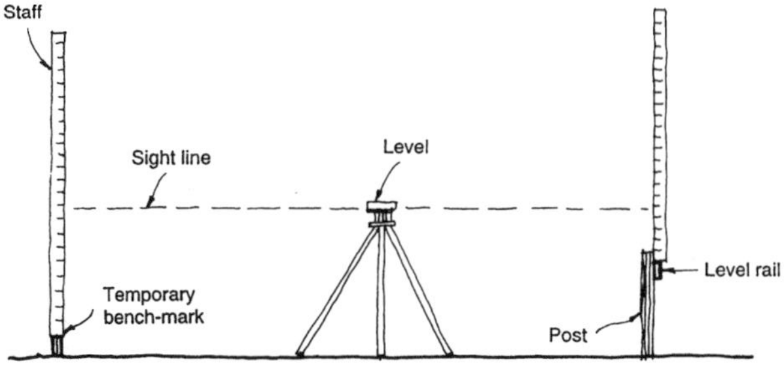
Staff
Sight line
Level
Temporary
bench-mark
Level rail
Post

Accurate survey

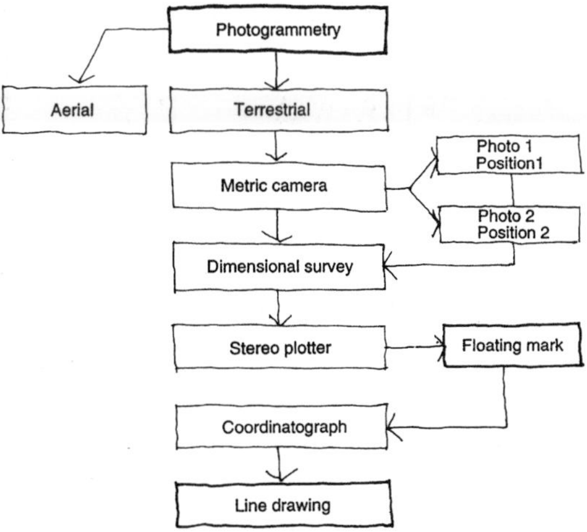
Photogrammetry
Aerial
Terrestrial
Metric camera
Photo 1
Position1
Photo 2
Position 2
Dimensional survey
Stereo plotter
Floating mark
Coordinatograph
Line drawing

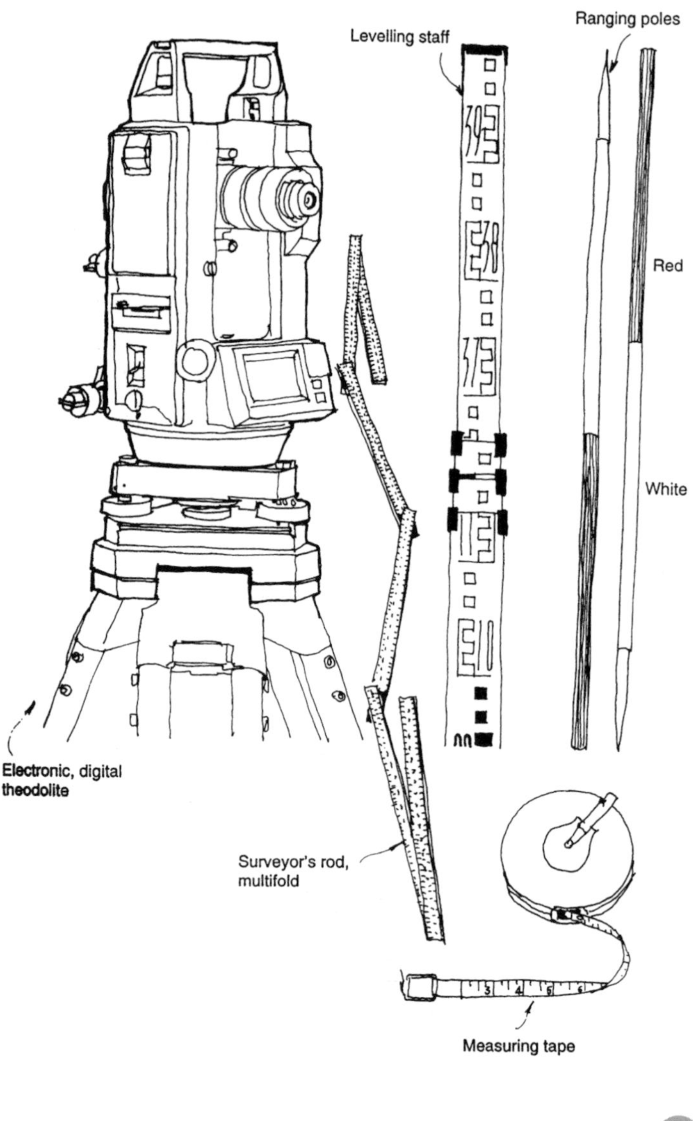
Ranging poles
Levelling staff
Red
White
Electronic, digital theodolite
Surveyor's rod, multifold
Measuring tape

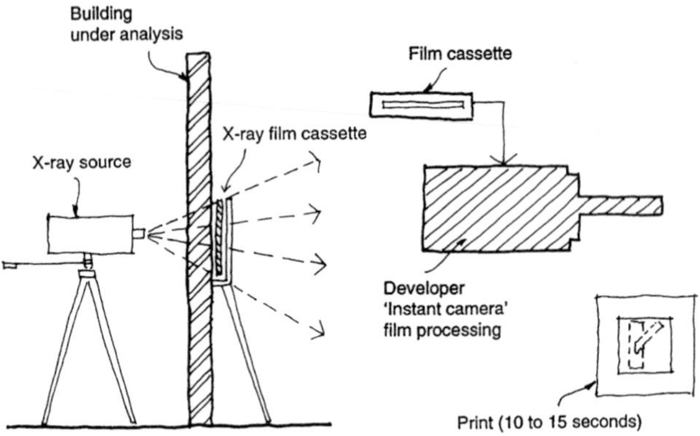
Building
under analysis
X-ray source
X-ray film cassette
Film cassette
Developer
'Instant camera'
film processing
Print (10 to 15 seconds)

Radiography

Magnetometry

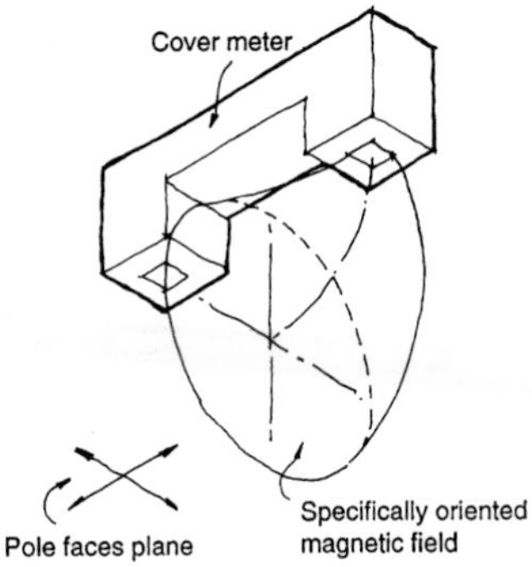
Cover meter
Pole faces plane
Specifically oriented
magnetic field

Microwave analysis

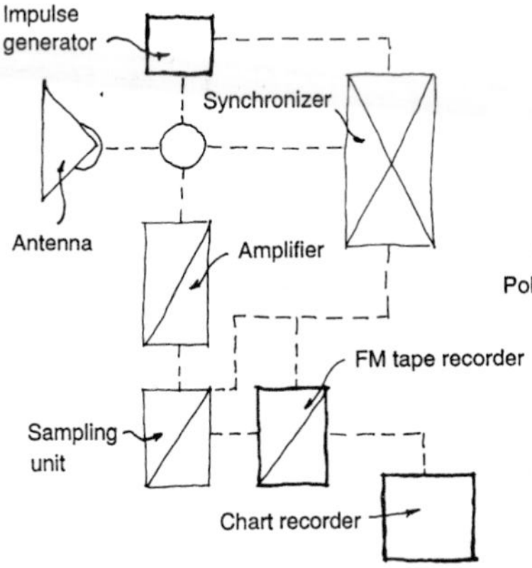
Impulse
generator
Synchronizer
Antenna
Amplifier
FM tape recorder
Sampling
unit
Chart recorder

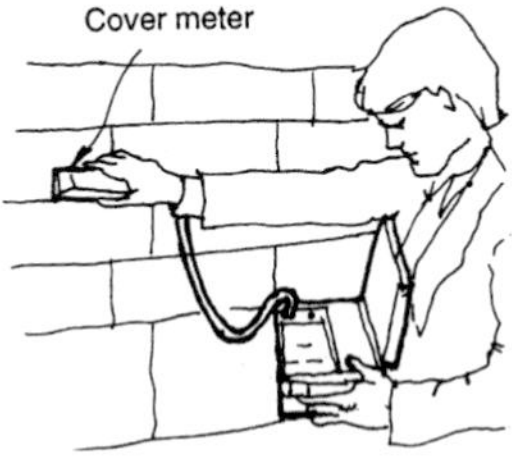
Cover meter

LVDT (Linear Variable Differential Transducers)

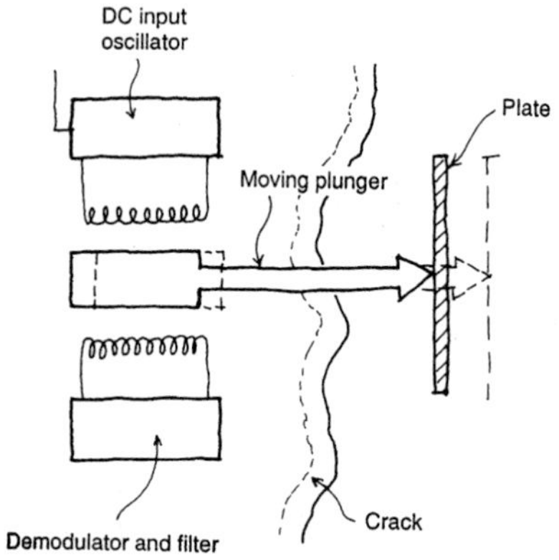

Transducer monitoring

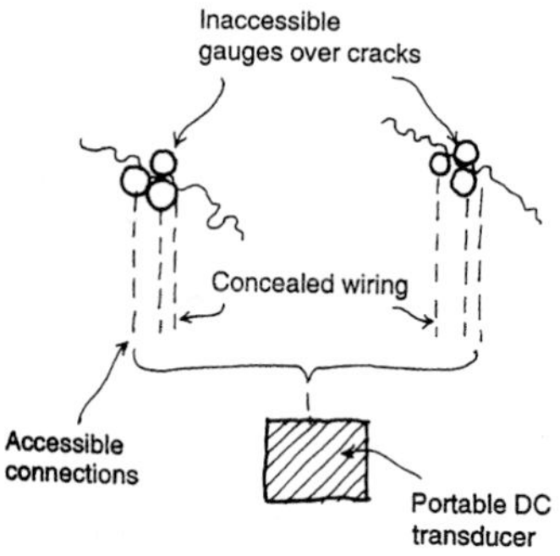

Portable read-out unit

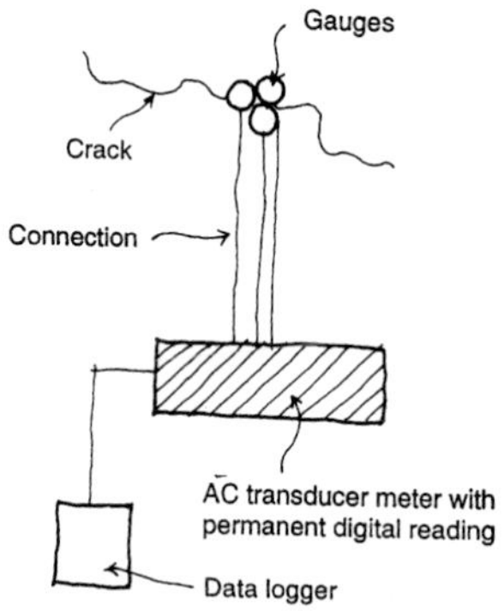

Monitoring with permanent transducer meter

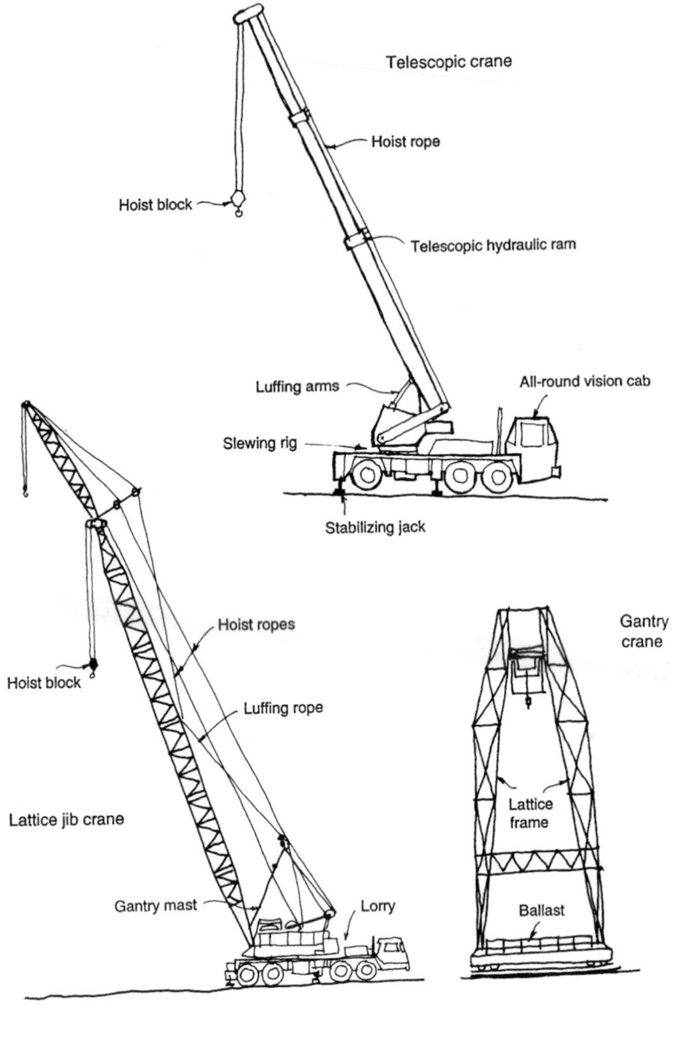
Telescopic crane
Hoist rope
Hoist block
Telescopic hydraulic ram
Luffing arms
All-round vision cab
Slewing rig
Stabilizing jack
Hoist ropes
Hoist block
Luffing rope
Lattice jib crane
Gantry mast
Lorry
Gantry crane
Lattice frame
Ballast

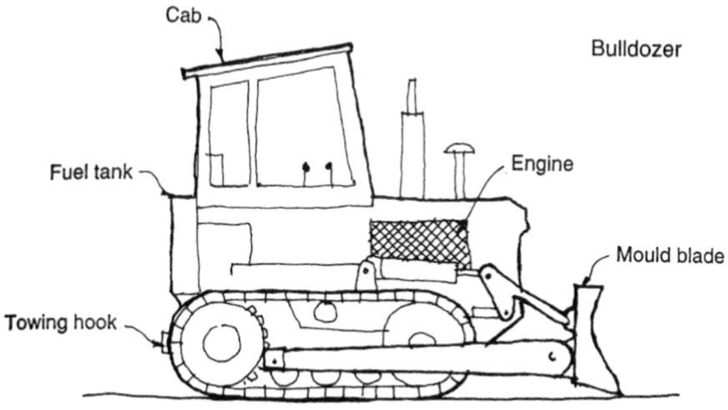

Bulldozer

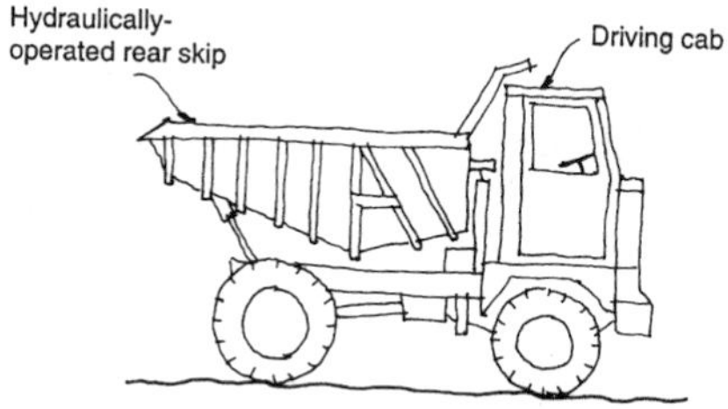

Highway dumper

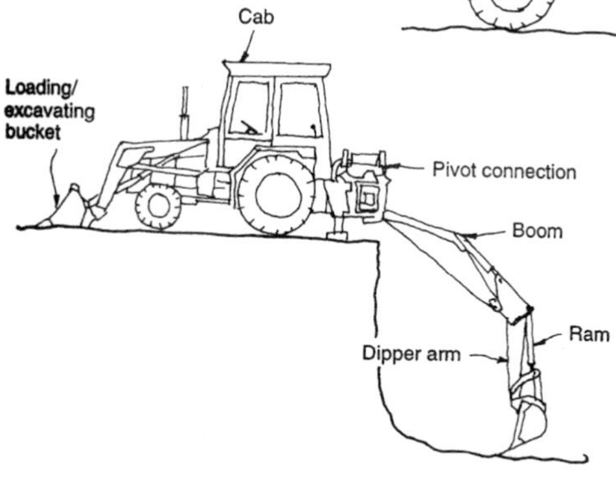

Multipurpose excavator

Scaffolding tower

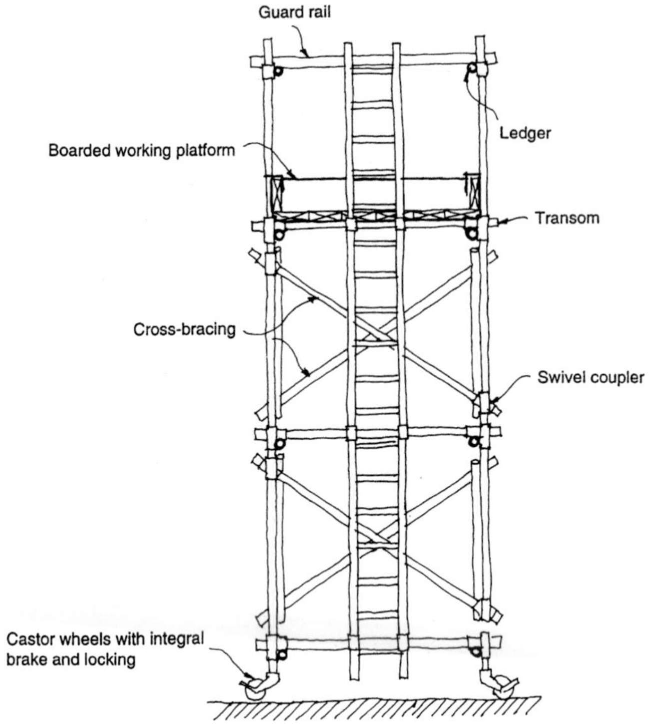
Guard rail
Ledger
Boarded working platform
Transom
Cross-bracing
Swivel coupler
Castor wheels with integral
brake and locking

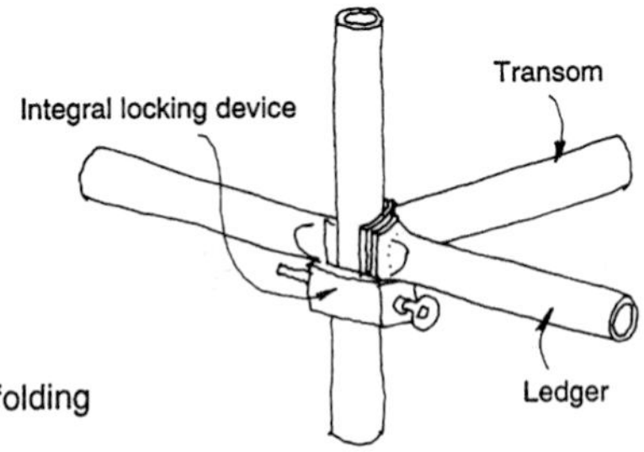
Transom
Integral locking device
Ledger

Patent scaffolding

Independent scaffold
(truss-out)

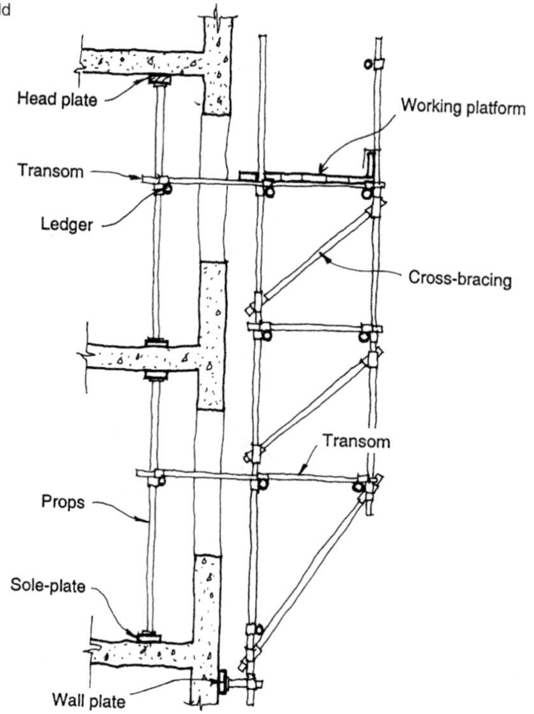

Scaffolding board

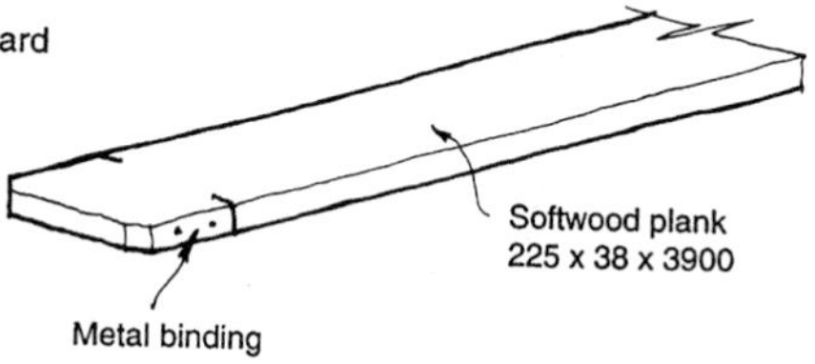

Dumper

Ready mix concrete truck

Tamping board vibrator

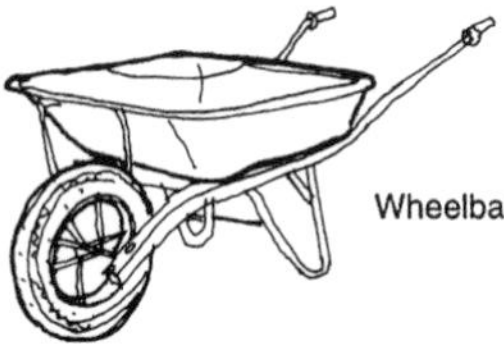

Wheelbarrow

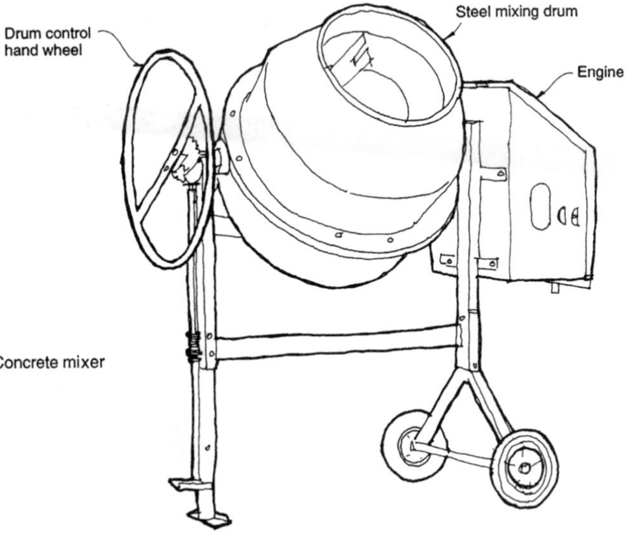

Concrete mixer

Generator set

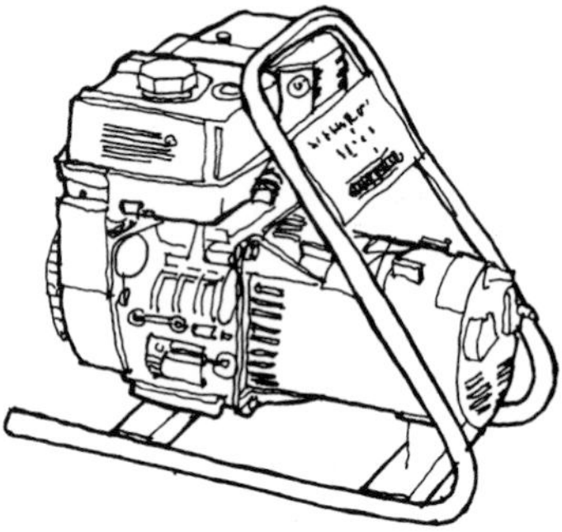

Rammer

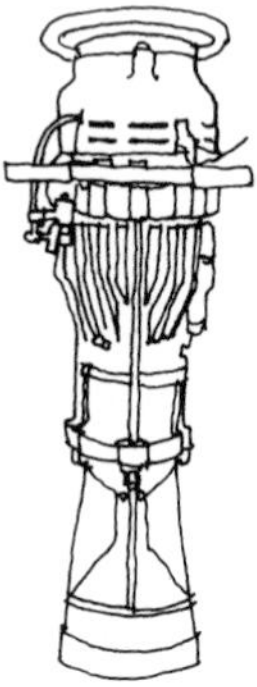

Compressor

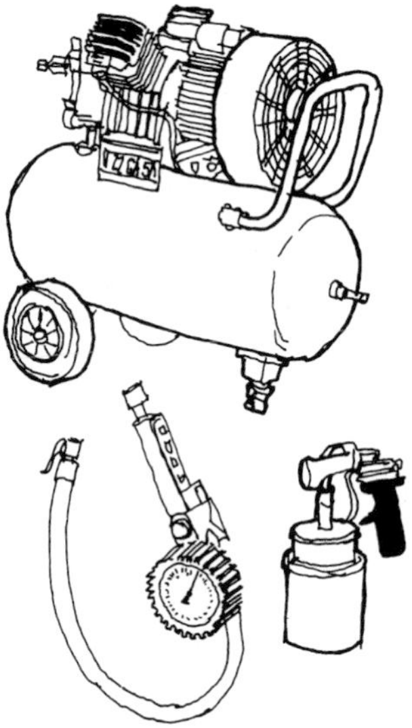

Pick

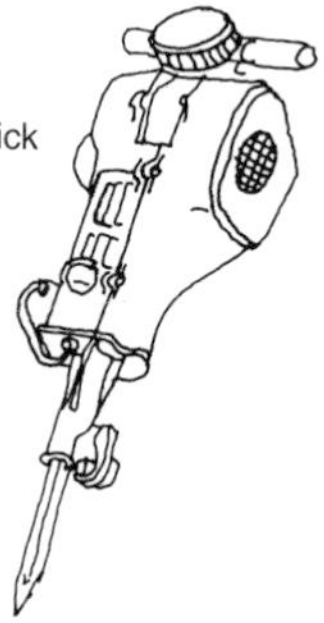

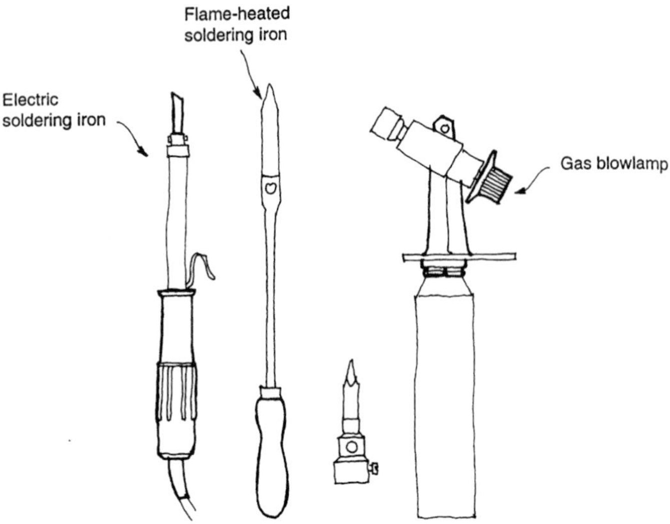
Flame-heated soldering iron
Electric soldering iron
Gas blowlamp

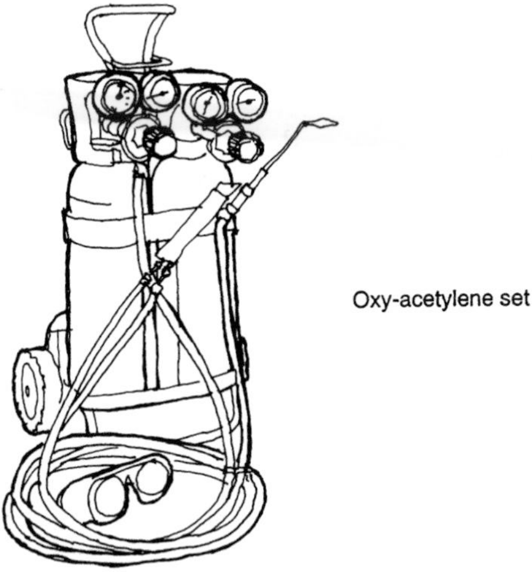
Oxy-acetylene set

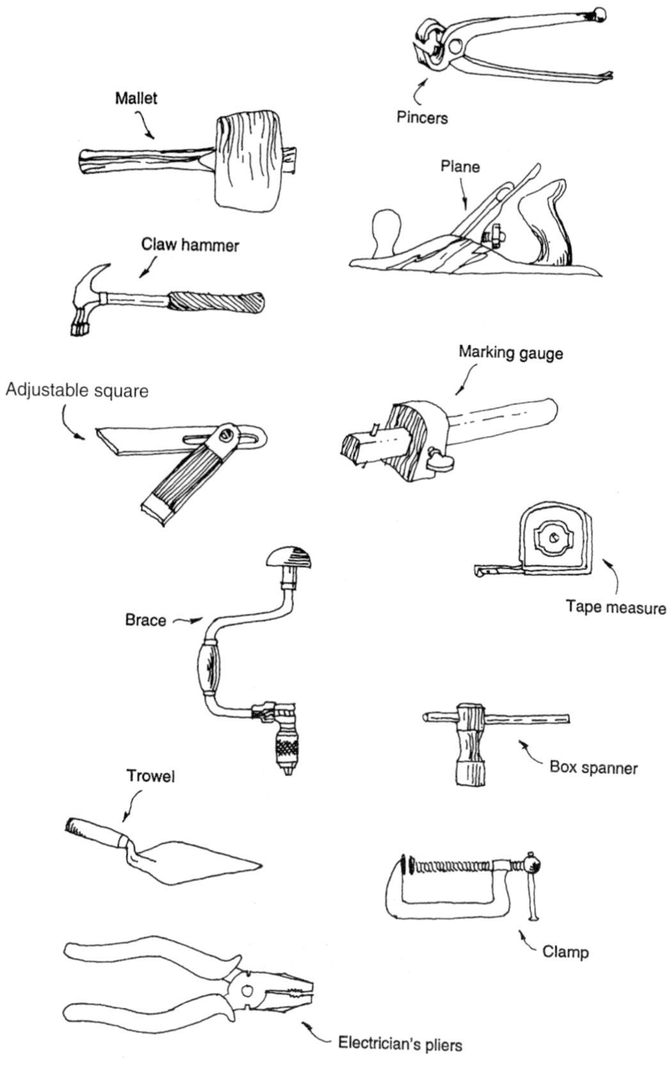
Mallet
Pincers
Plane
Claw hammer
Marking gauge
Adjustable square
Tape measure
Brace
Box spanner
Trowel
Clamp
Electrician's pliers

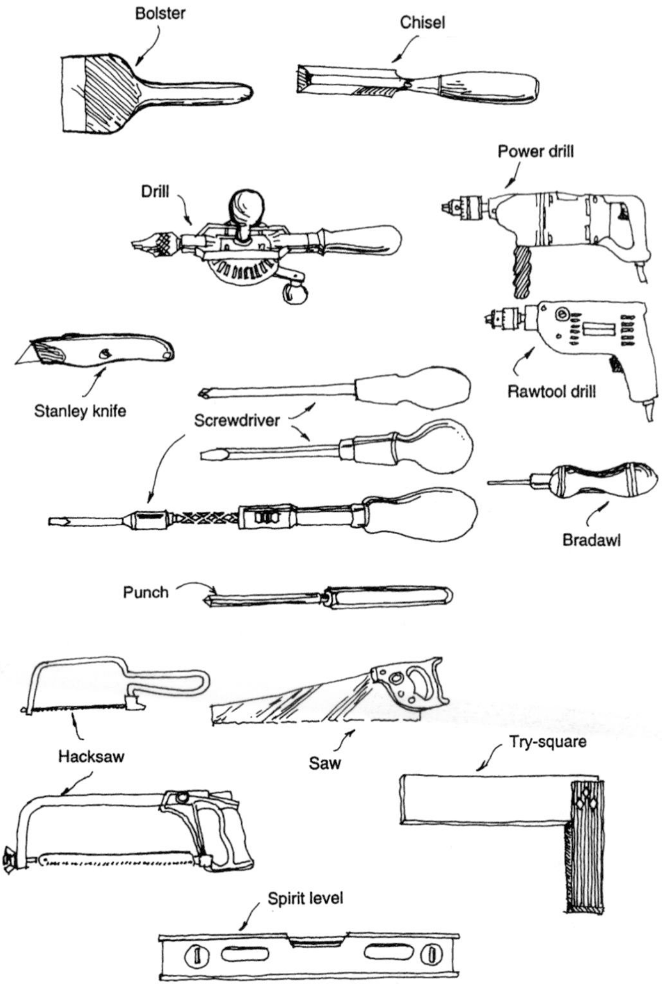
Bolster
Chisel
Power drill
Drill
Stanley knife
Rawtool drill
Screwdriver
Bradawl
Punch
Hacksaw
Saw
Try-square
Spirit level

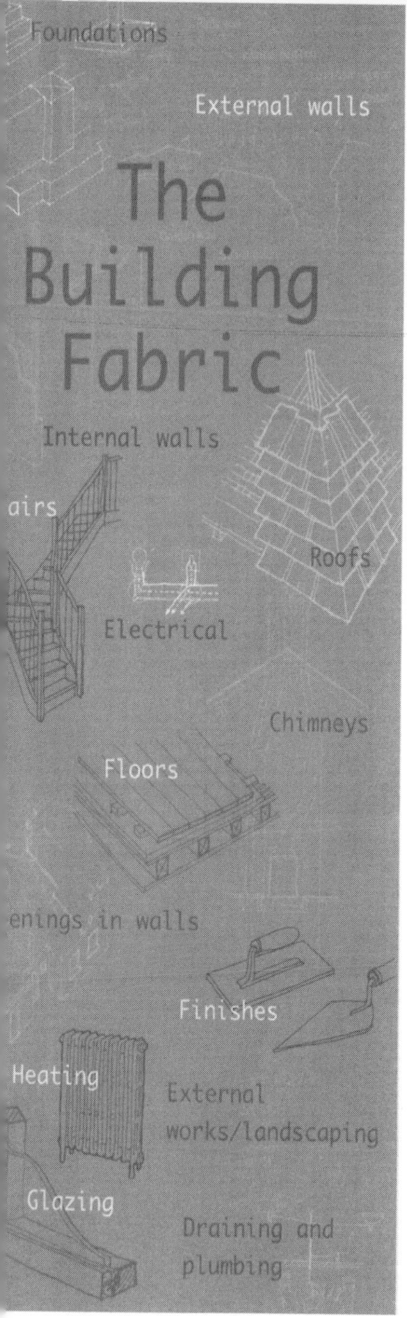
Foundations
External walls
The Building Fabric
Internal walls
airs
Roofs
Electrical
Chimneys
Floors
enings in walls
Finishes
Heating
External works/landscaping
Glazing
Draining and plumbing

V. The Building Fabric

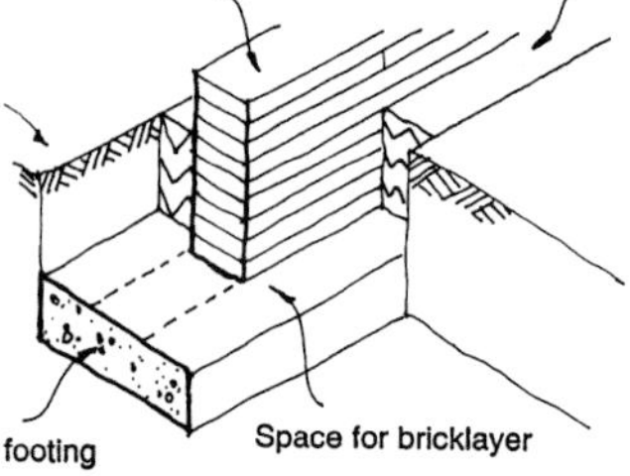

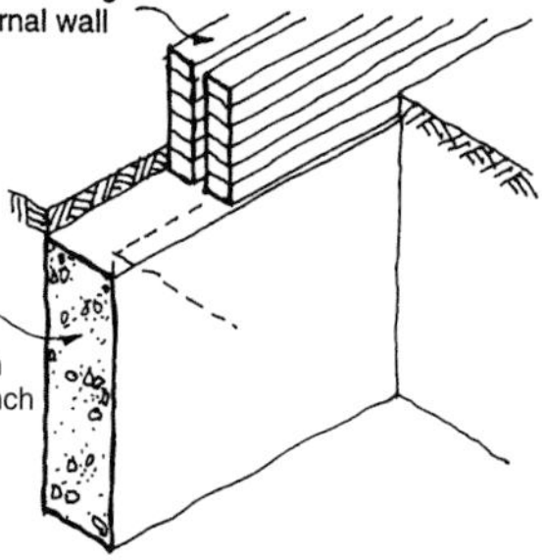

Trench fill foundations (deep)

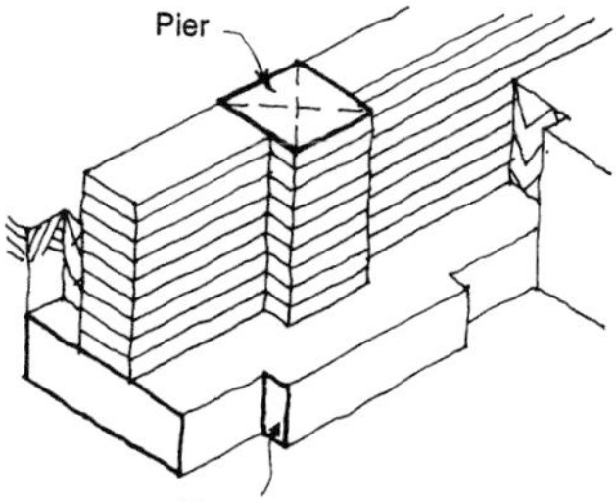

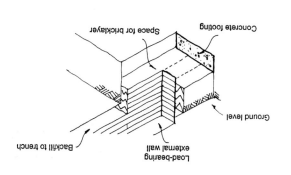

Wall pier foundations

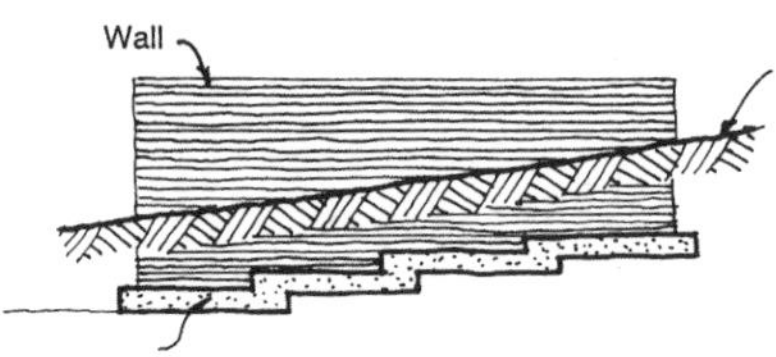
Wall
Ground line
Stepped concrete foundation

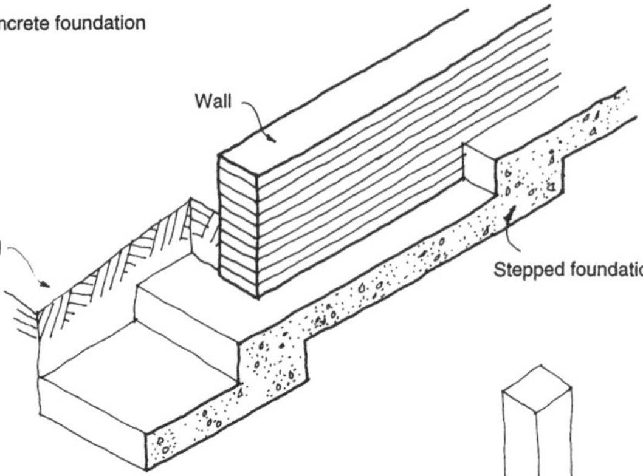
Wall
Ground
Stepped foundation

Detail

Isolated pad foundations

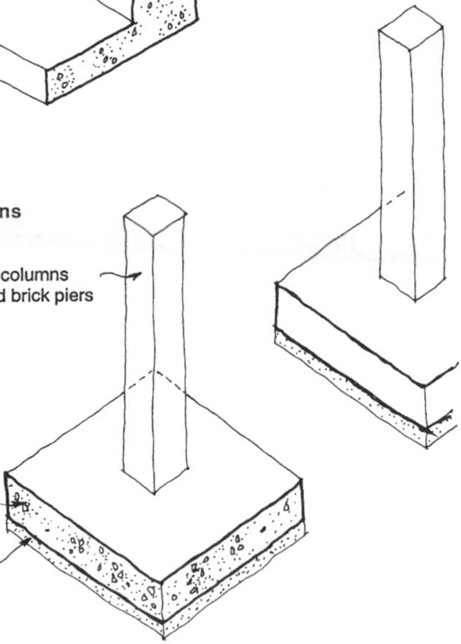
Concrete columns
or isolated brick piers
Concrete base
Blinding

(for columns closely spaced)

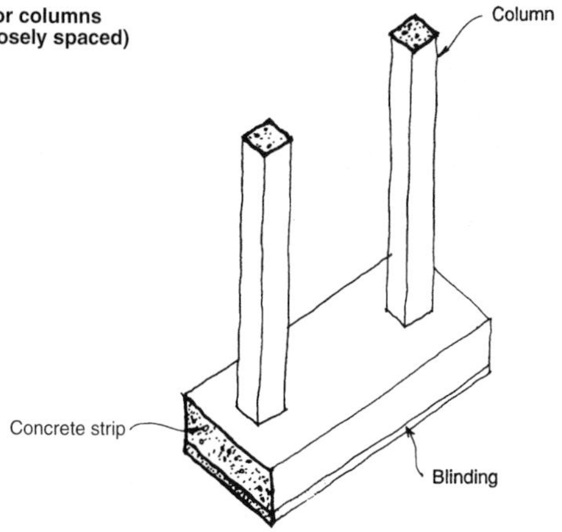
Column
Concrete strip
Blinding

Raft foundations

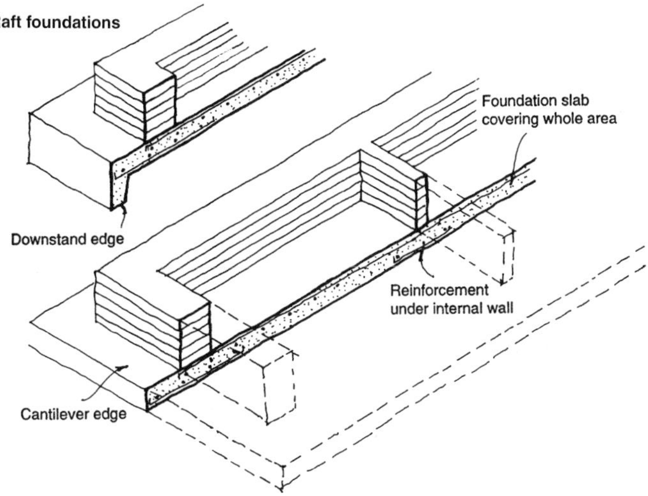
Foundation slab
covering whole area
Downstand edge
Reinforcement
under internal wall
Cantilever edge

Short-bored piles

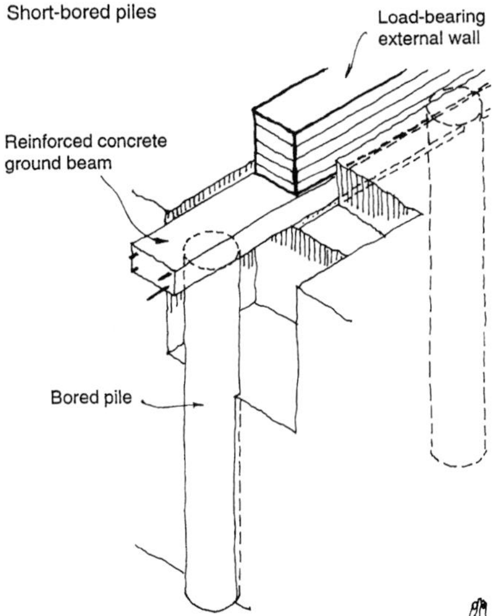
Load-bearing
external wall
Reinforced concrete
ground beam
Bored pile

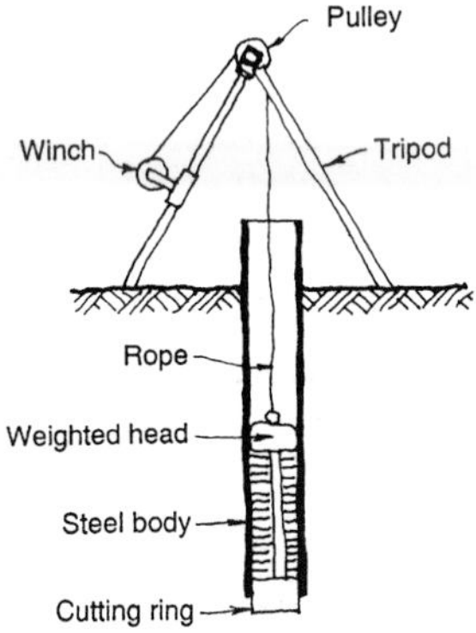
Pulley
Winch
Tripod
Rope
Weighted head
Steel body
Cutting ring

Percussion bored

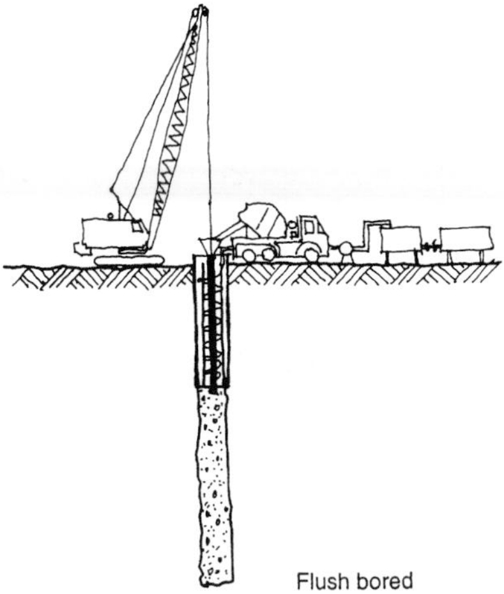

Flush bored

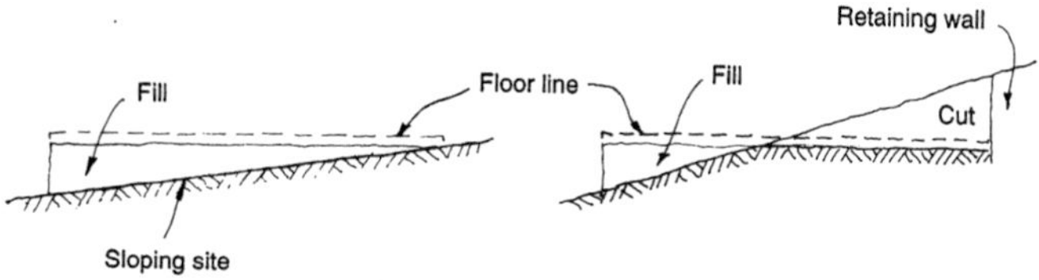

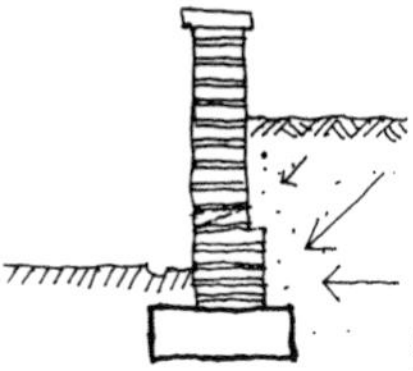

Small retaining wall

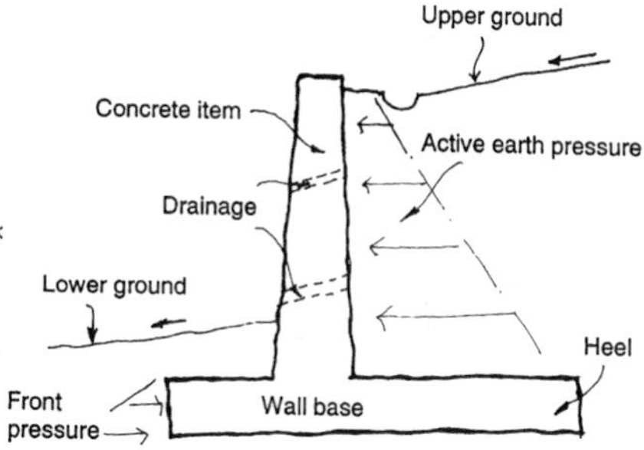

Medium height retaining wall

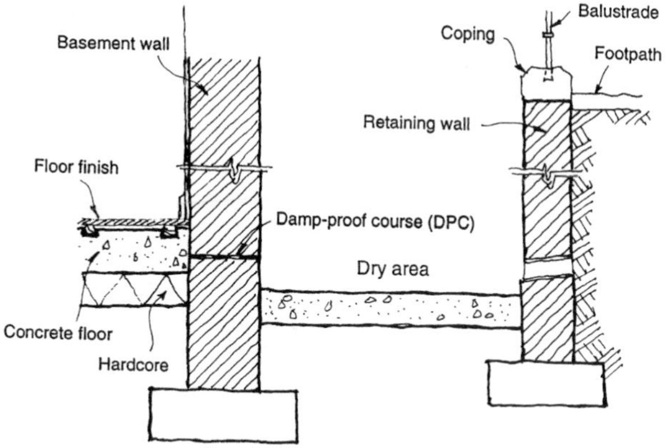

Steel column foundations

Steel grillage foundation

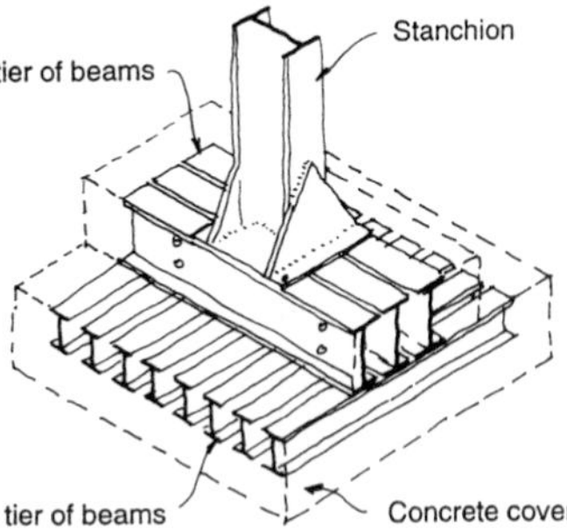

Multiple steel
column foundation

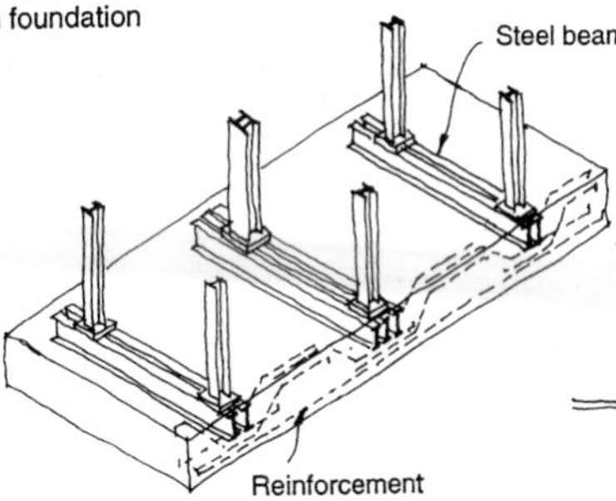

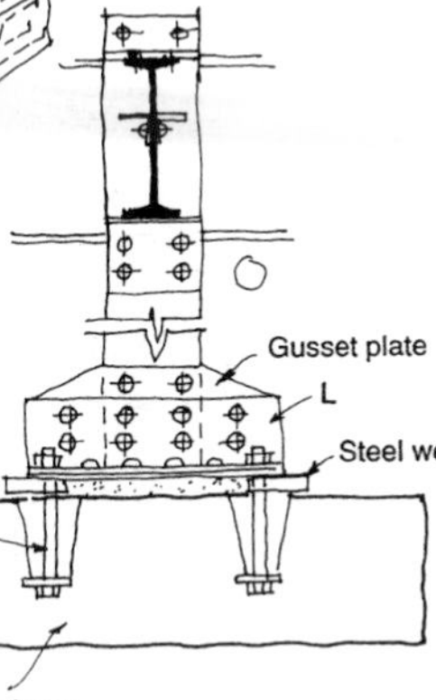

Shoring and underpinning

Asymmetrical flying shore

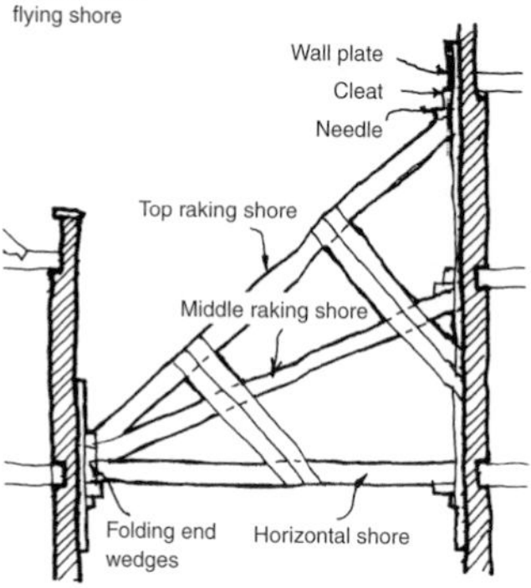

Multiple raking shore

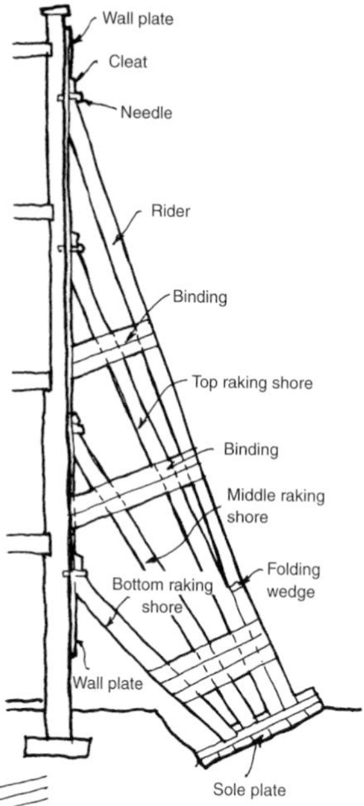

Underpinning

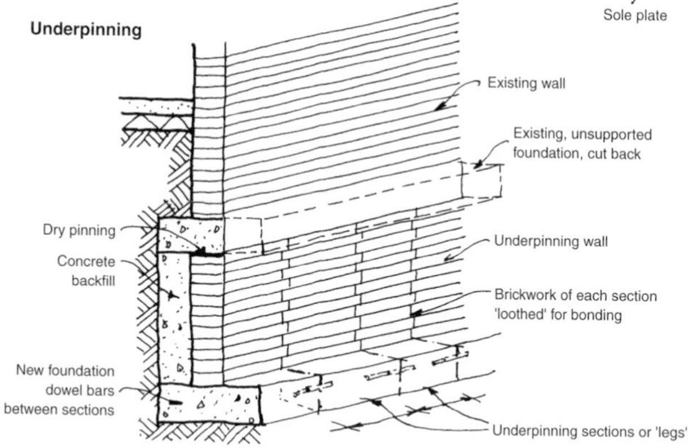

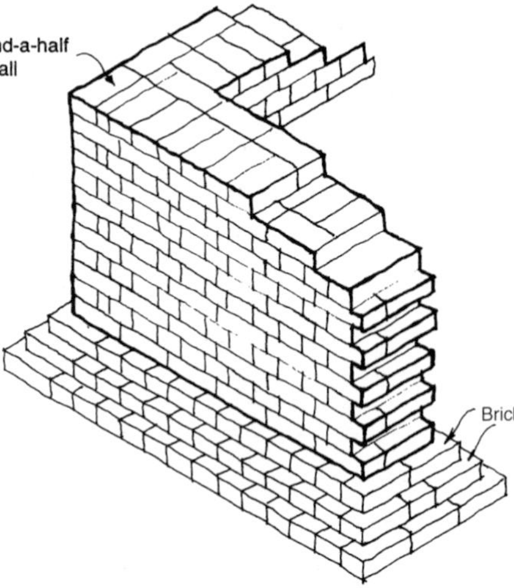
One-and-a-half
brick wall
Brick footings

Solid brick wall

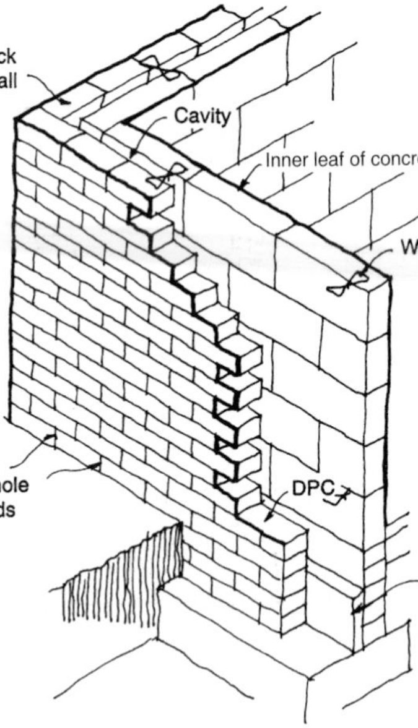
Half-brick
outer wall
Cavity
Inner leaf of concrete blocks
Wall tie
Weep-hole
perpends
DPC
Concrete
cavity fill

Cavity wall

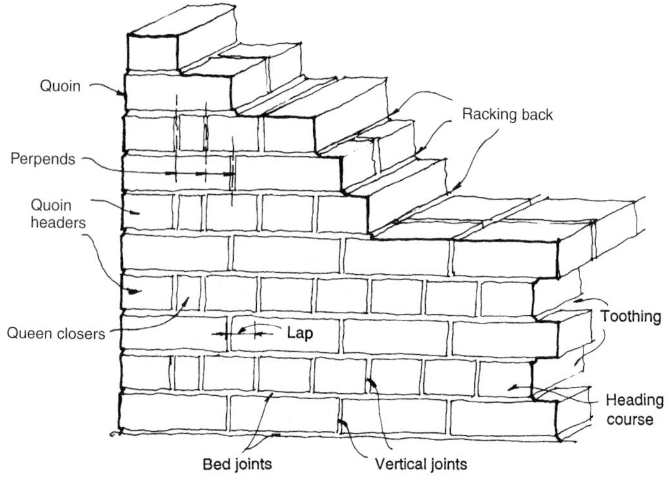
Quoin
Perpends
Quoin headers
Queen closers
Racking back
Lap
Toothing
Heading course
Bed joints
Vertical joints

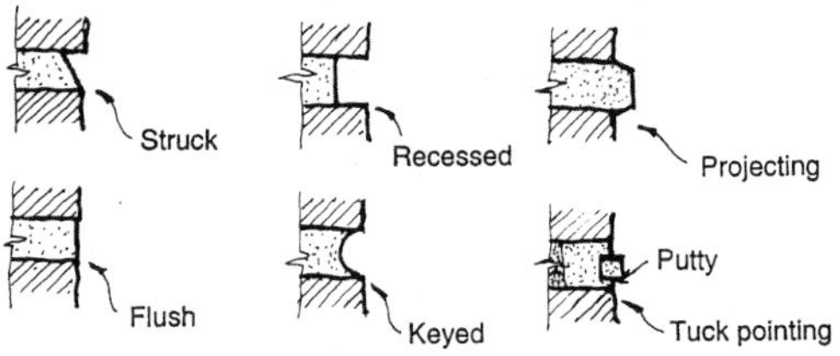
Struck
Recessed
Projecting
Flush
Keyed
Putty
Tuck pointing

Type of joints

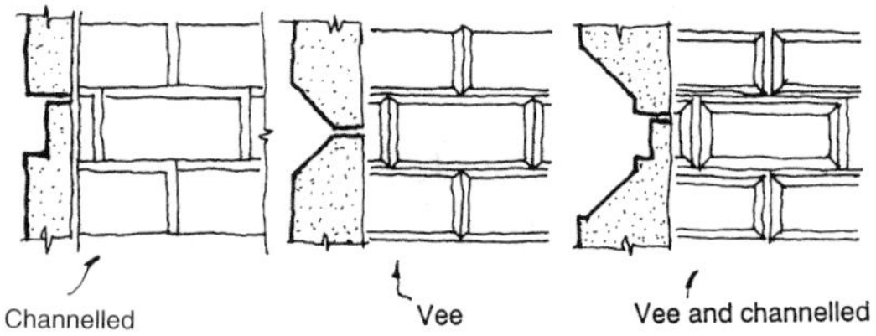
Channelled
Vee
Vee and channelled

Rusticated joints

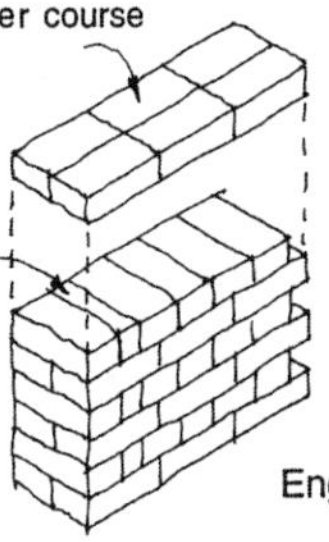

English bond

English cross-bond

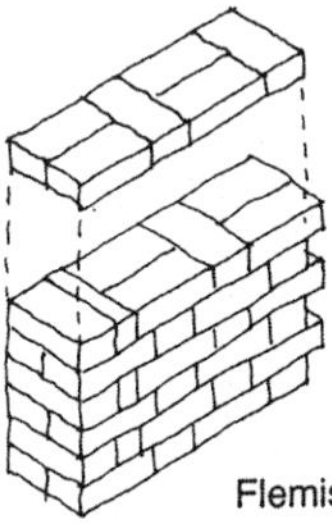
Flemish bond

Rat-trap bond

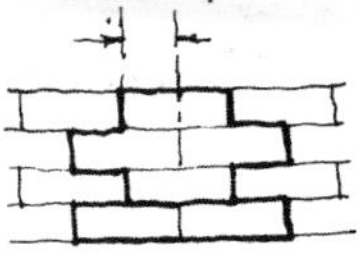

Half bonding

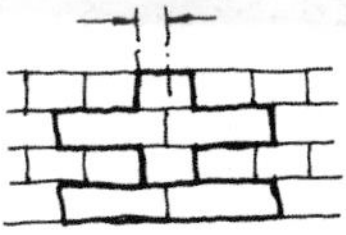

Quarter bonding

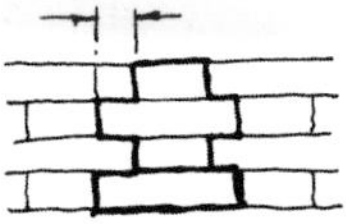

Third bonding

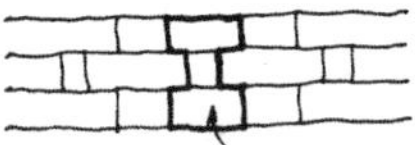

Broken bond

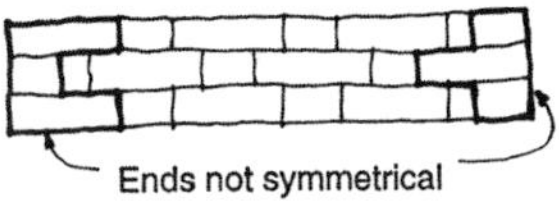

Reverse bond

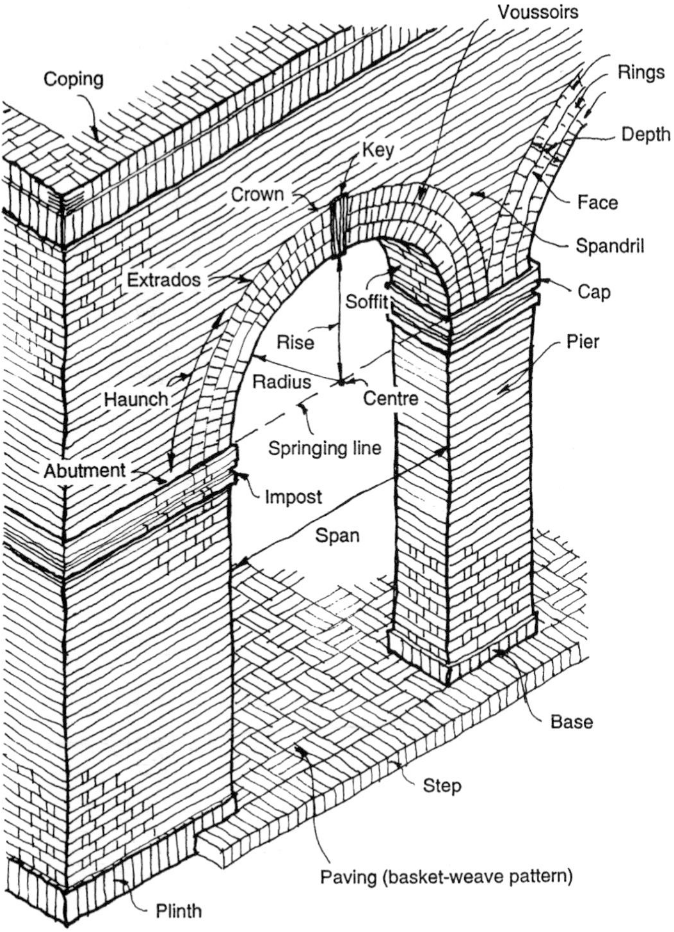
Voussoirs
Coping
Rings
Depth
Key
Crown
Face
Spandril
Extrados
Cap
Soffit
Rise
Pier
Radius
Haunch
Centre
Springing line
Abutment
Impost
Span
Base
Step
Paving (basket-weave pattern)
Plinth

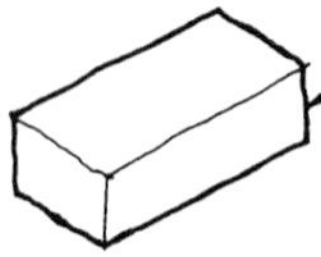

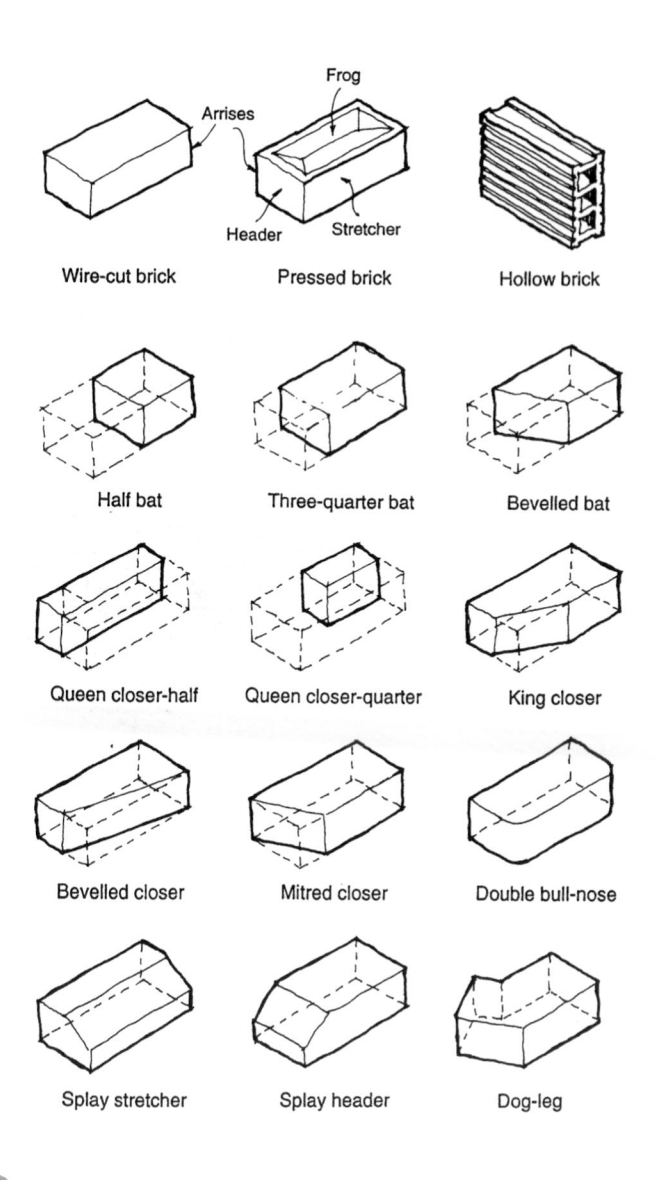

Wire-cut brick

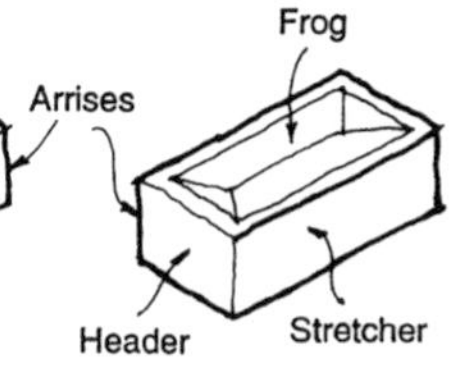

Pressed brick

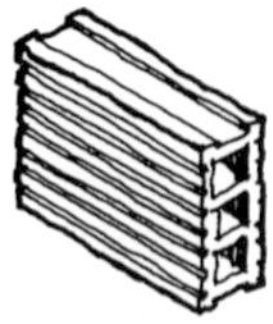

Hollow brick

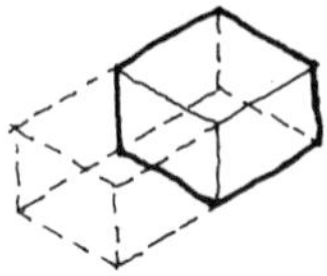

Half bat

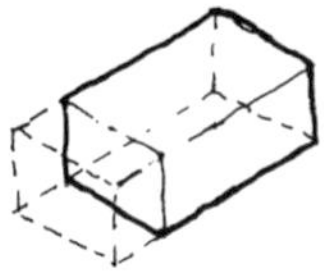

Three-quarter bat

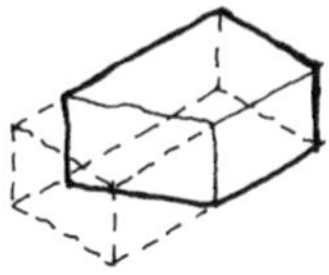

Bevelled bat

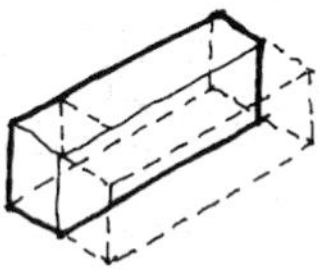

Queen closer-half

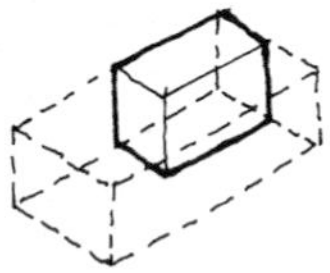

Queen closer-quarter

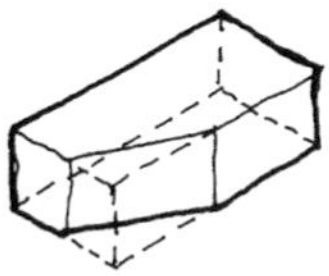

King closer

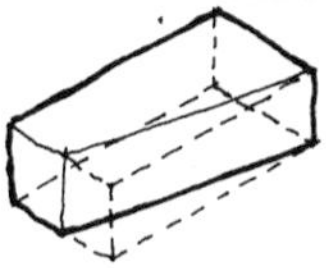

Bevelled closer

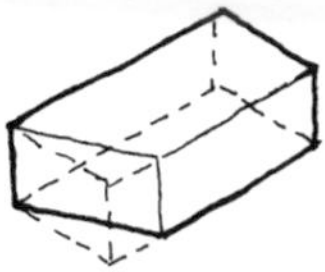

Mitred closer

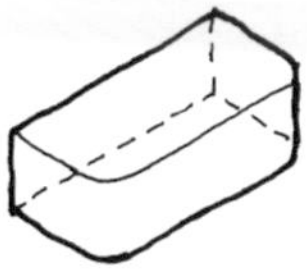

Double bull-nose

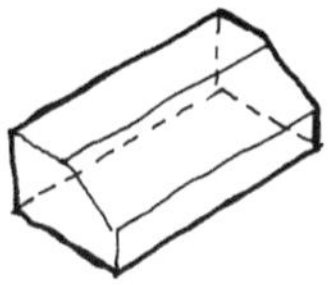

Splay stretcher

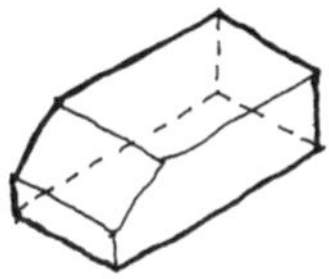

Splay header

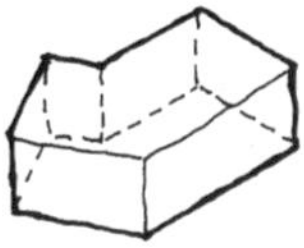

Dog-leg

Efflorescence

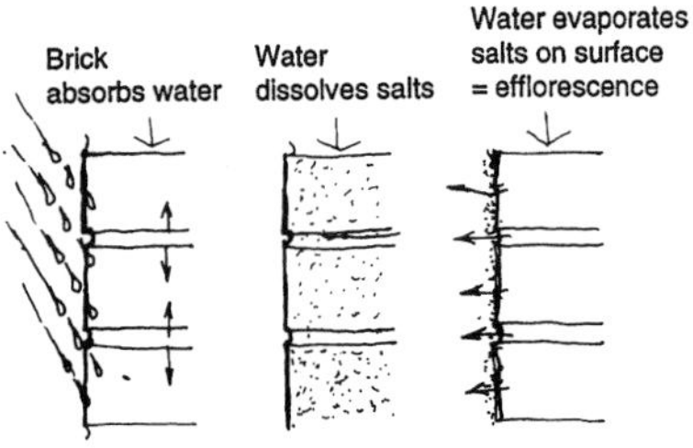
Brick
absorbs water
Water
dissolves salts
Water evaporates
salts on surface
= efflorescence

Efflorescence stains
and surface decay

Condensation

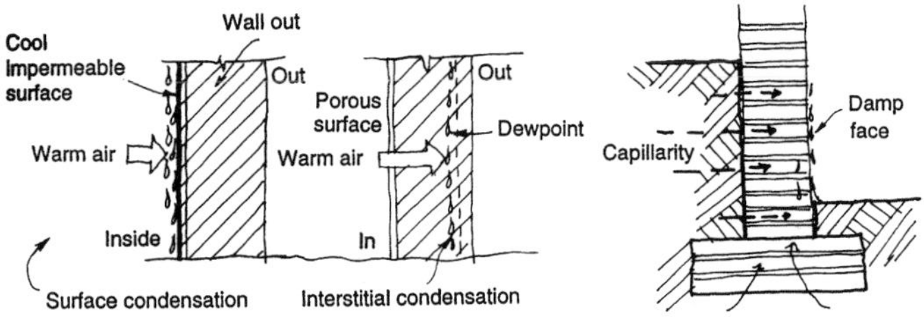
Wall out
Cool
Impermeable
surface
Out
Warm air
Inside
Surface condensation
Porous
surface
Out
Dewpoint
Warm air
In
Interstitial condensation
Damp
face
Capillarity

Section through face of limestone quarry

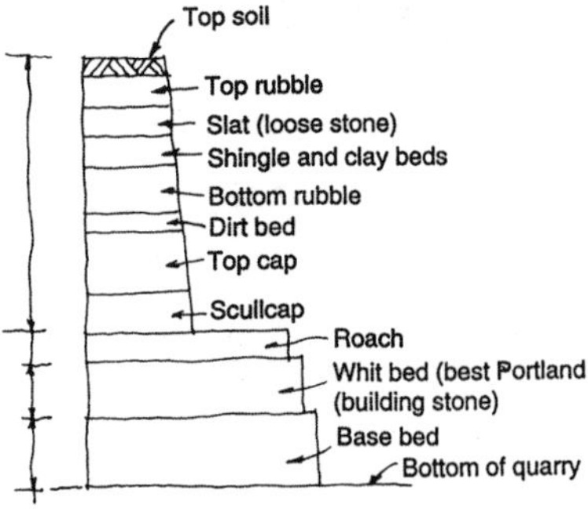

Surface finishes

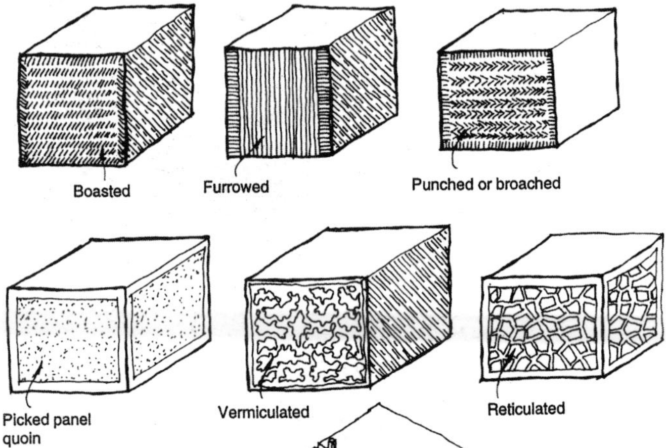

Coping A stone

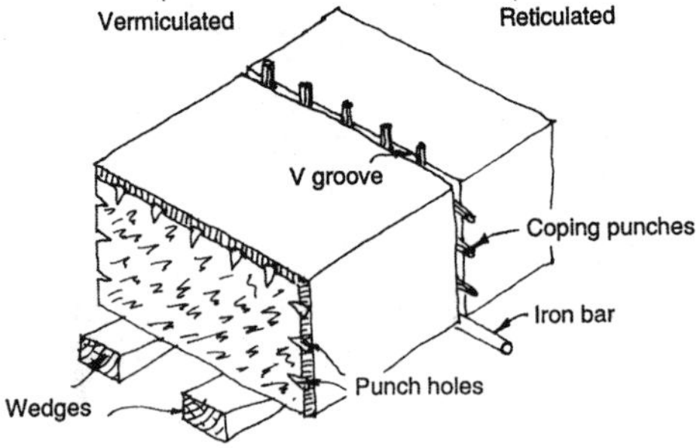

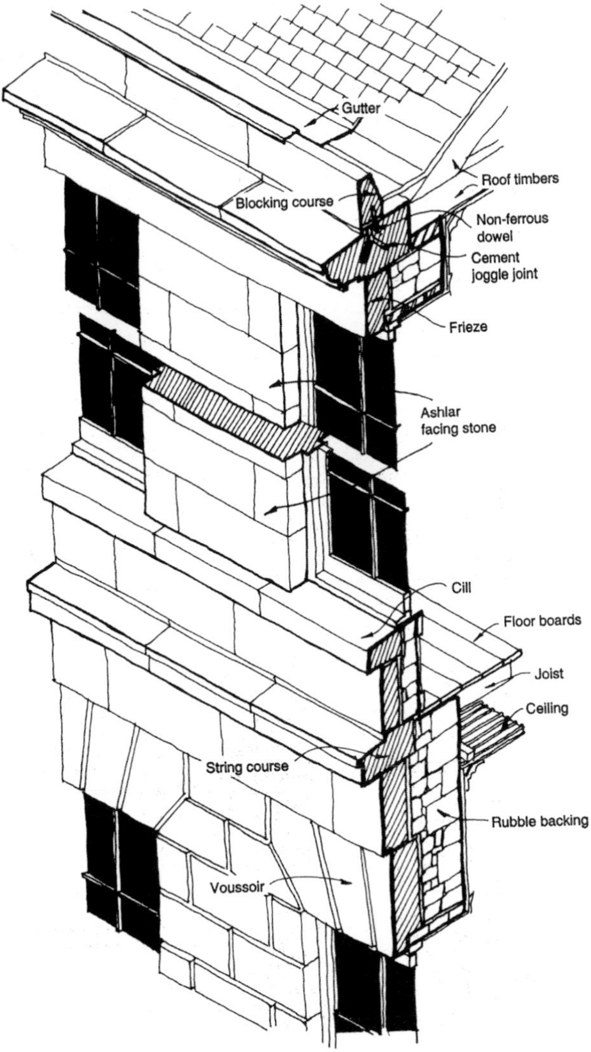
Gutter
Roof timbers
Blocking course
Non-ferrous dowel
Cement joggle joint
Frieze
Ashlar facing stone
Cill
Floor boards
Joist
Ceiling
String course
Rubble backing
Voussoir

Rubble work

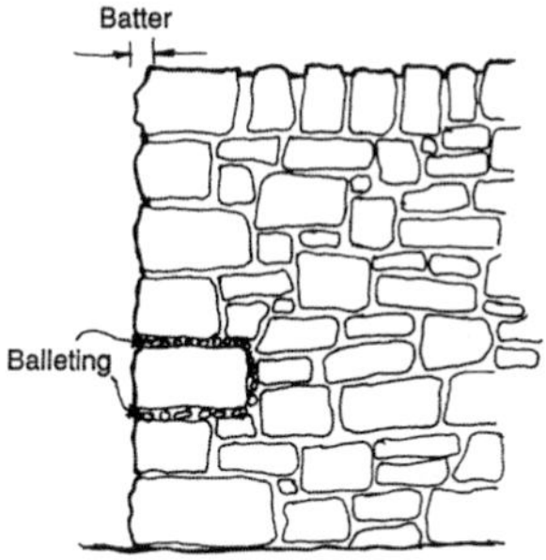

Random rubble uncoursed

Random rubble built to courses

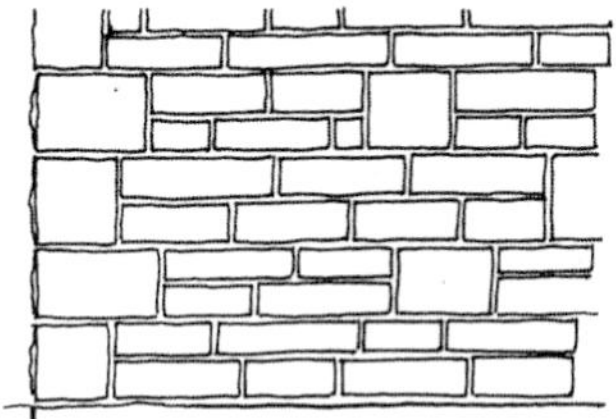

Squared rubble coursed

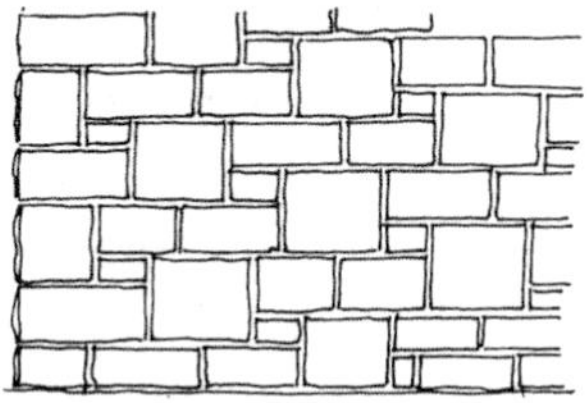

Squared rubble (snecked)

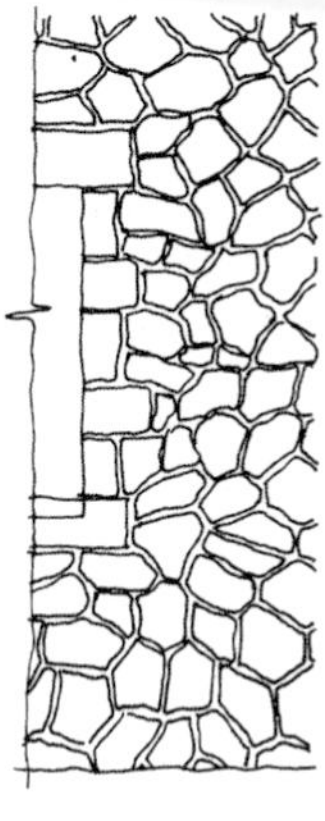

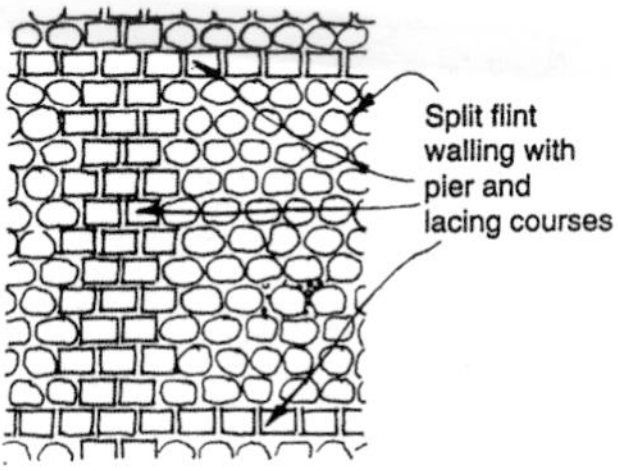

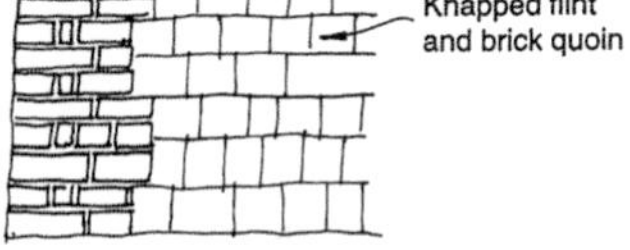

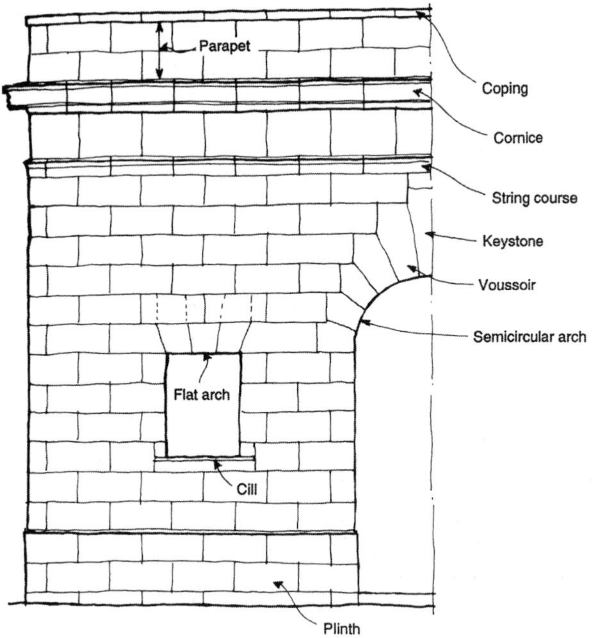

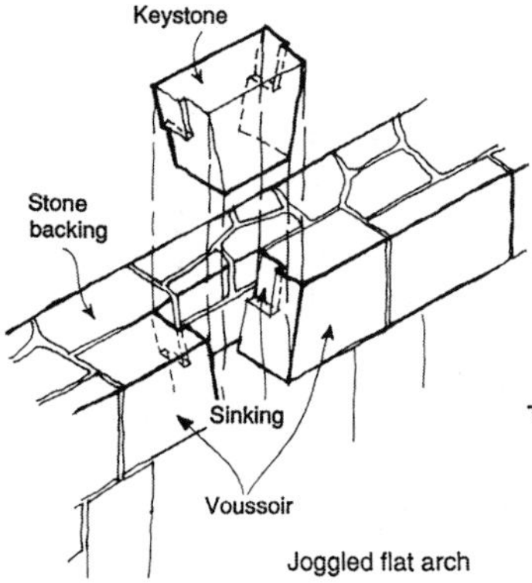

Joggled flat arch

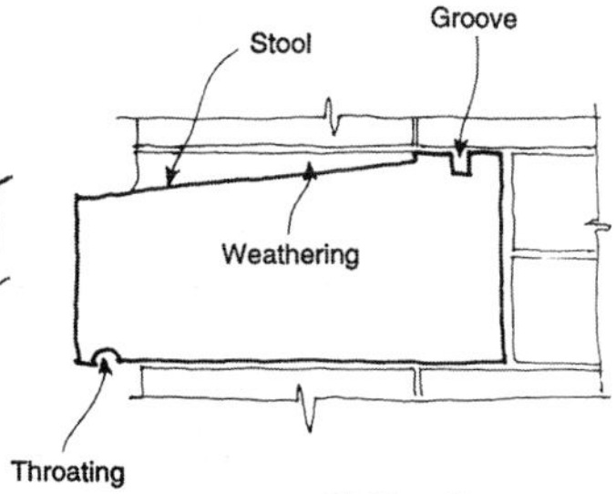

Window cill

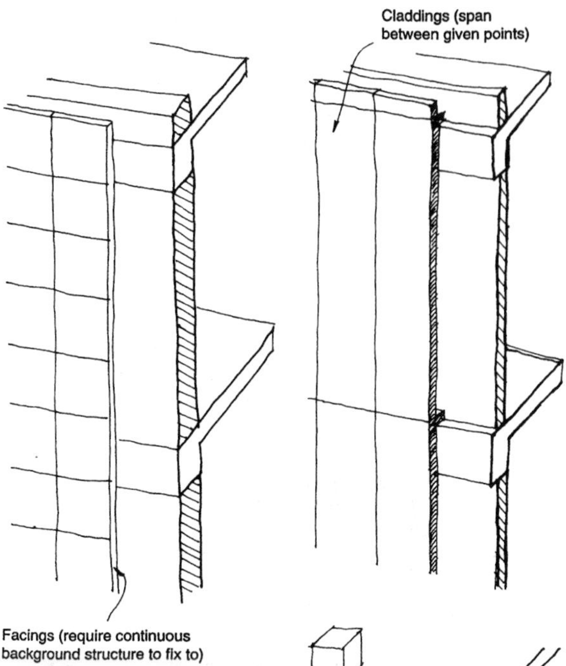

Cladding types:
Precast concrete slabs, curtain walling, profiled sheeting

Facings:
brick, stone, concrete slabs, tiling, weatherboarding

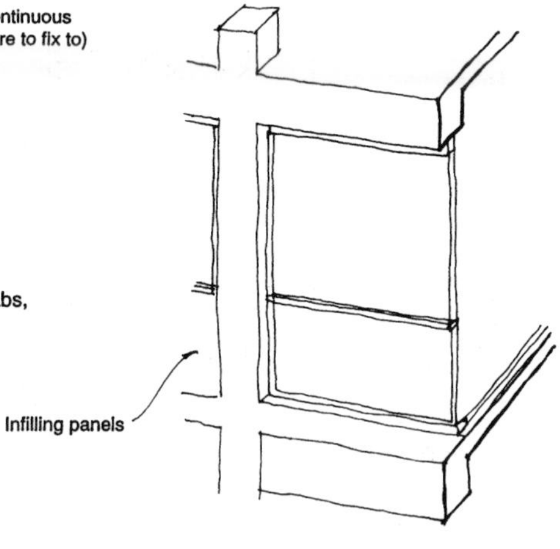

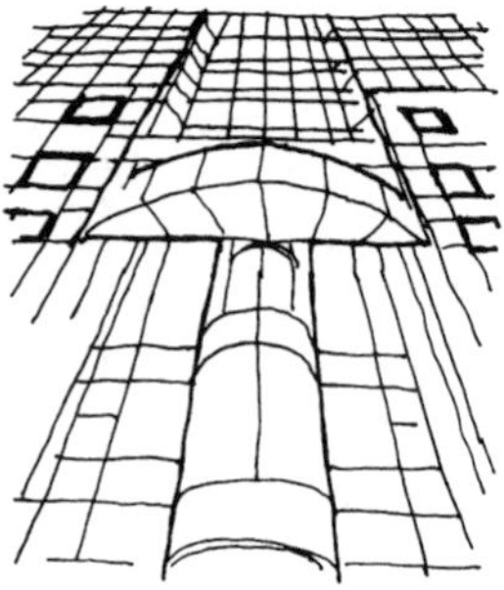

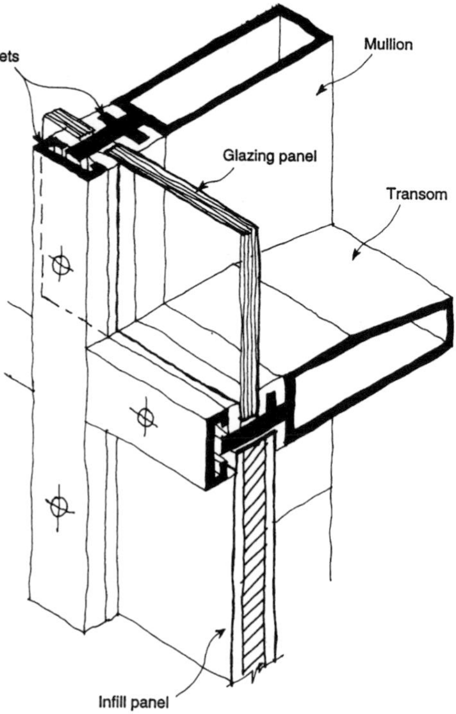

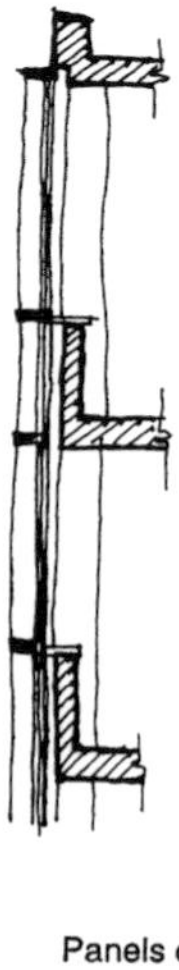

Panels cover structure

(glass fibre reinforced cement)

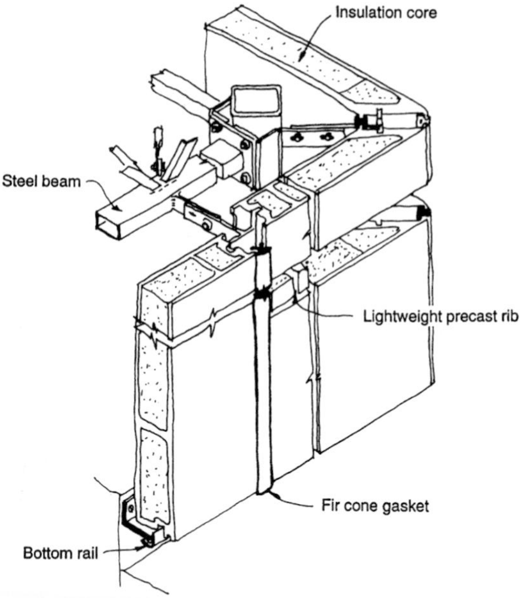

GRC component joints

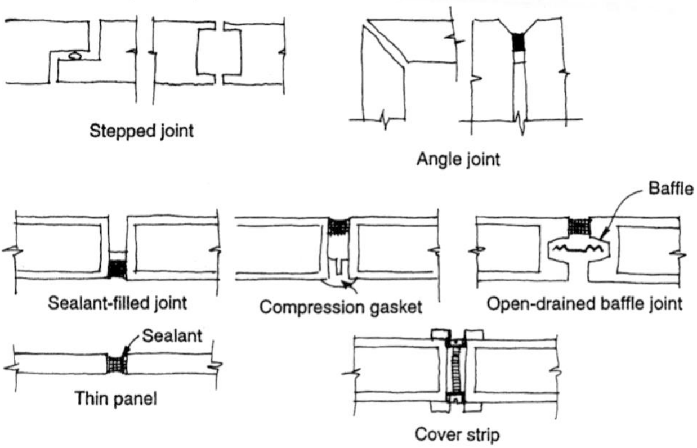

Lightweight – timber stud
Non-loadbearing

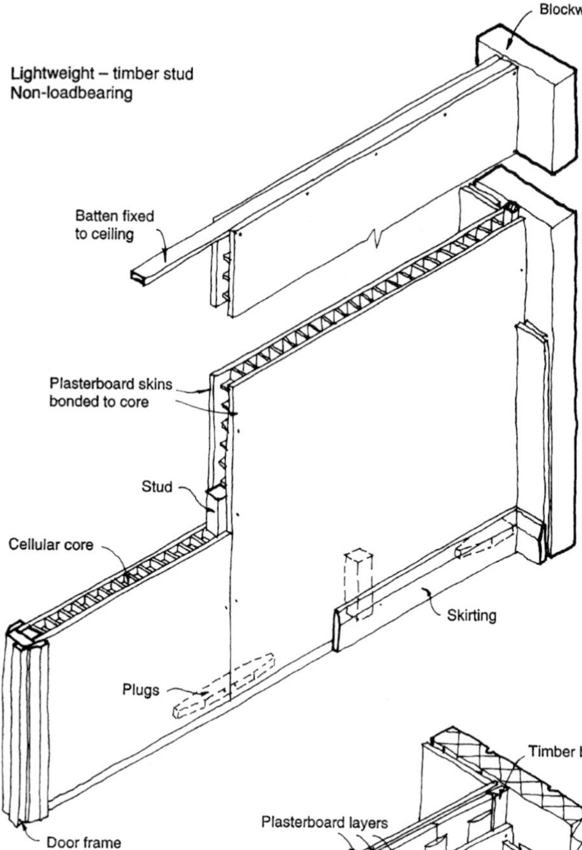

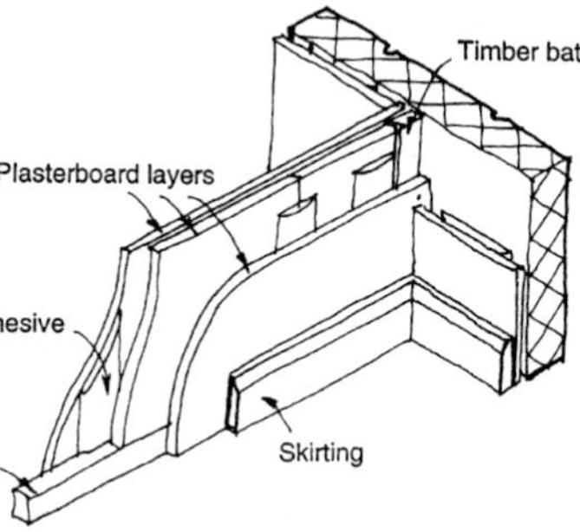

Laminated partition

Internal – Metal stud
Non-loadbearing

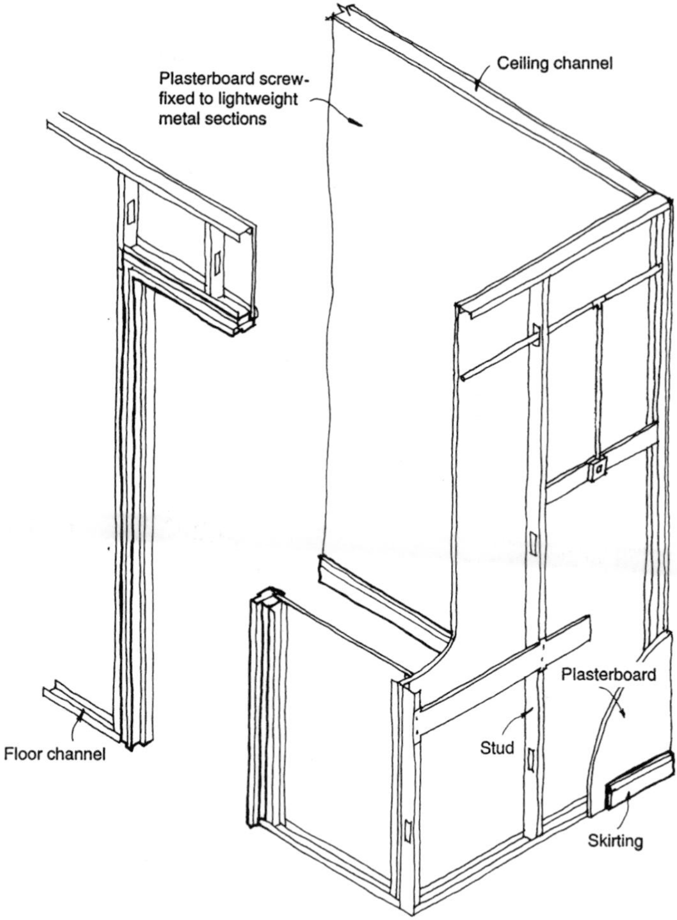
Plasterboard screw-
fixed to lightweight
metal sections
Ceiling channel
Floor channel
Plasterboard
Stud
Skirting

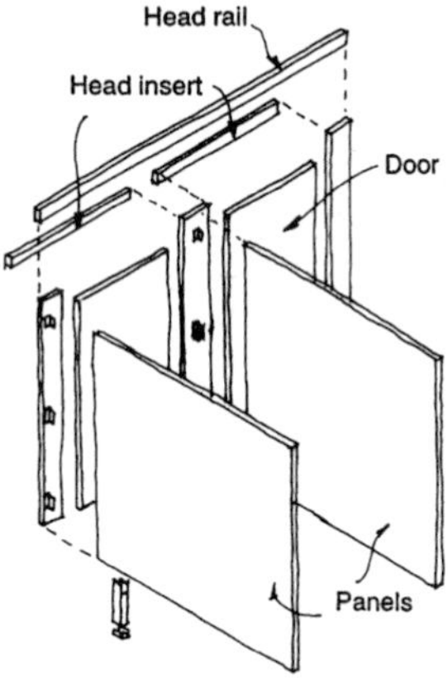

Proprietary cubicles

Demountable partitioning system

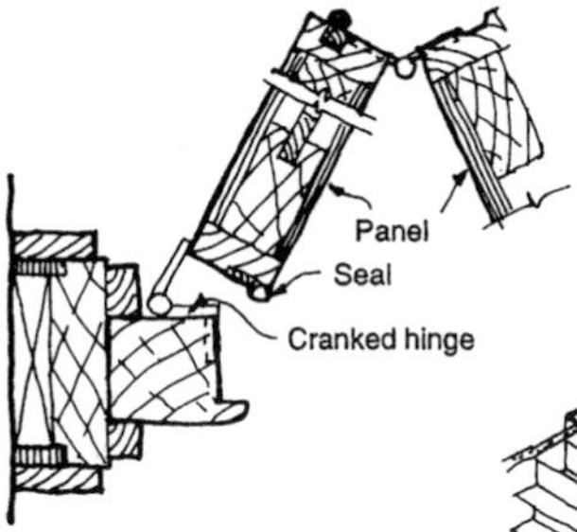

Sliding–folding hinged partition

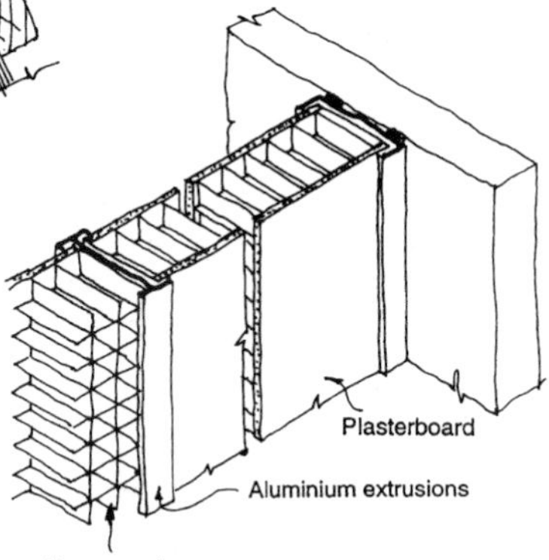

Frame and panel partition system

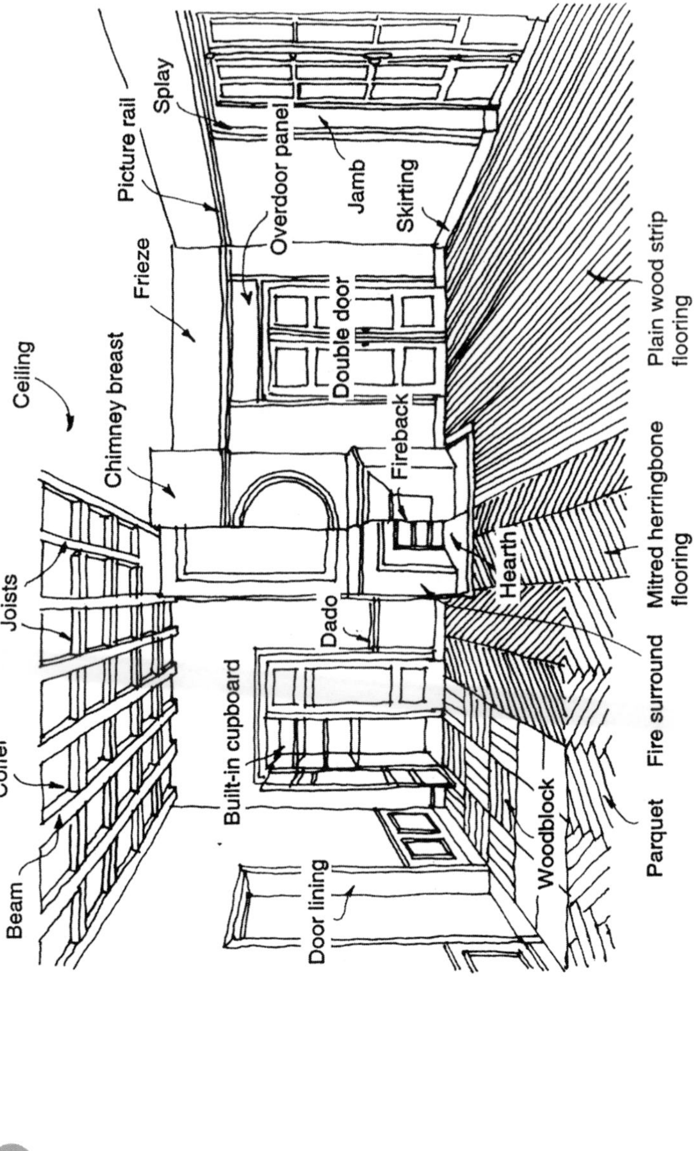
Ceiling
Chimney breast
Frieze
Picture rail
Splay
Overdoor panel
Jamb
Skirting
Double door
Fireback
Hearth
Dado
Built-in cupboard
Joists
Coffer
Beam
Door lining
Woodblock
Parquet
Fire surround
Mitred herringbone flooring
Plain wood strip flooring

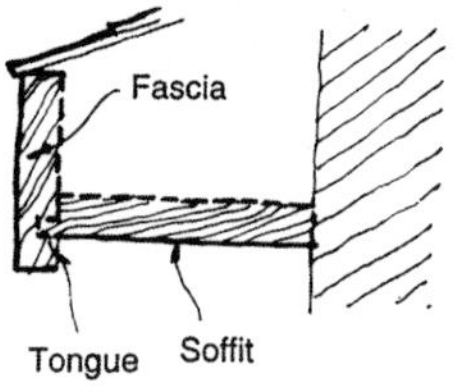
Fascia
Tongue
Soffit

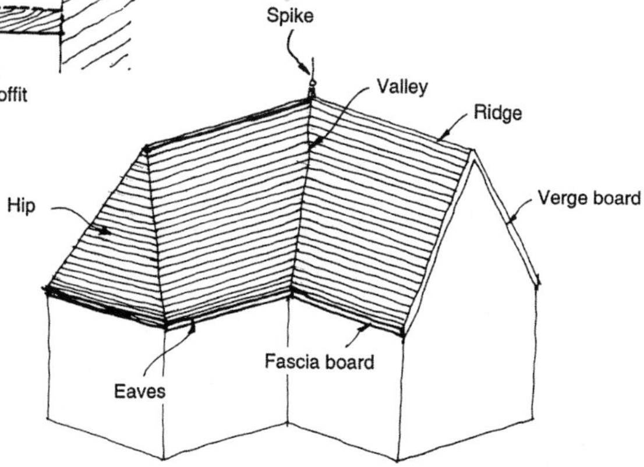
Spike
Valley
Ridge
Hip
Verge board
Fascia board
Eaves

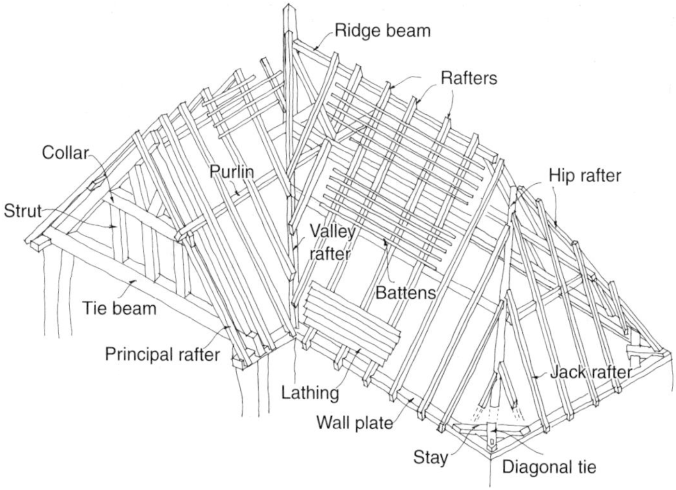
Ridge beam
Rafters
Collar
Purlin
Hip rafter
Strut
Valley rafter
Battens
Tie beam
Principal rafter
Jack rafter
Lathing
Wall plate
Stay
Diagonal tie

Types of pitched roofs

Monopitch

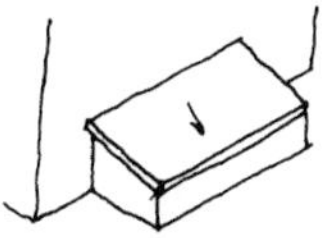

Lean-to

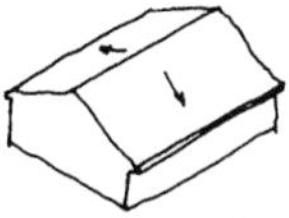

Ridge

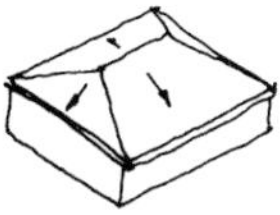

Hipped

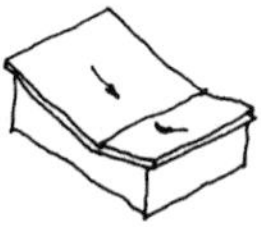

Butterfly

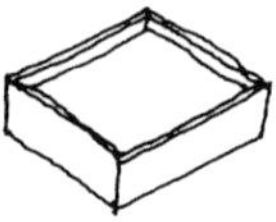

Flat roof

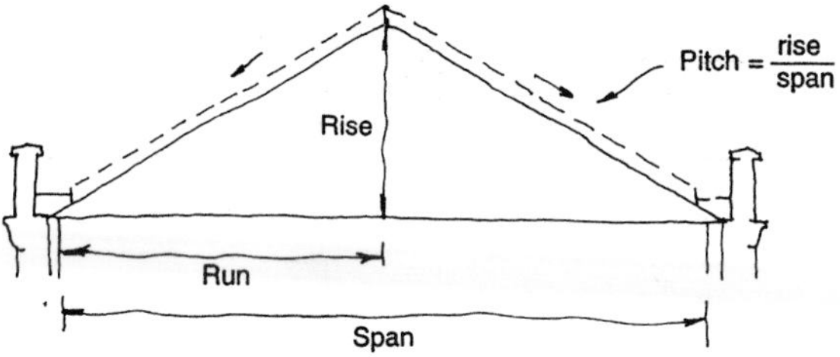

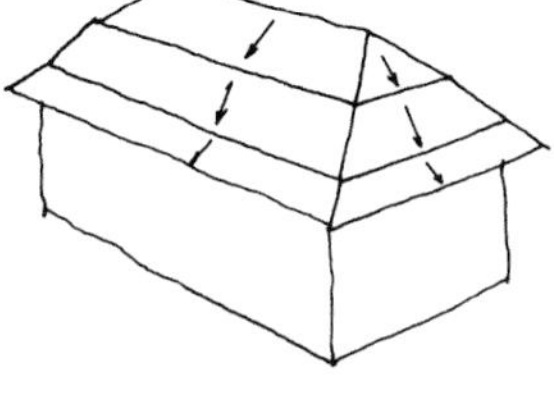

Mansarded

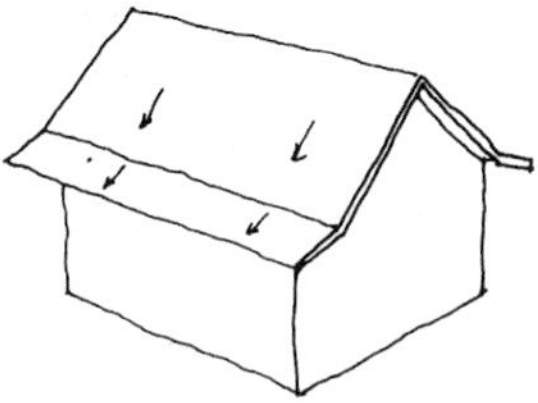

Splayed (sprocketed)

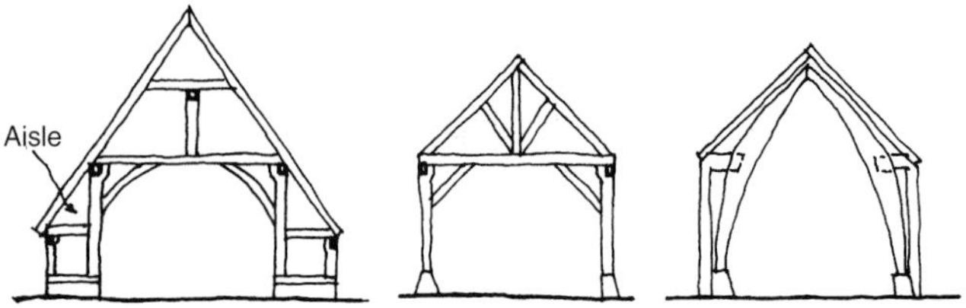

Crown post King post Cruck

Collar beam and braces Scissor beam

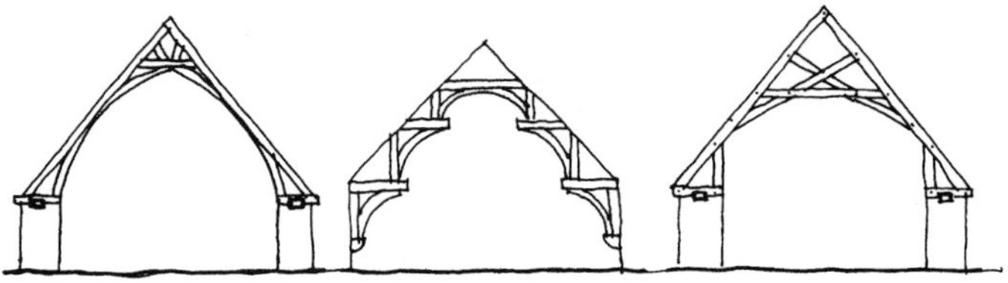

Arch-braced Double hammer-beam Collar with scissor-beams

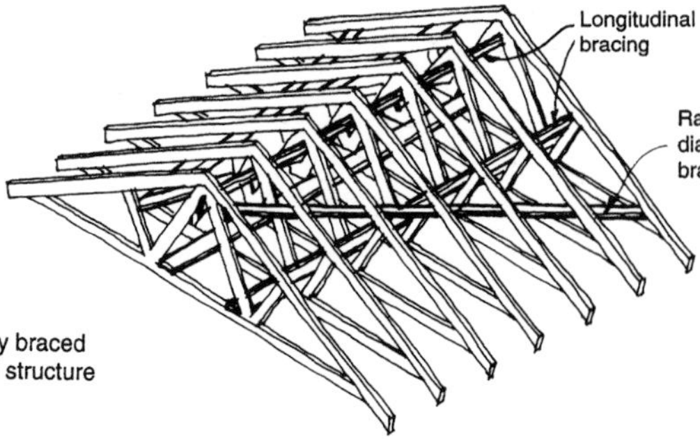

Fully braced
roof structure

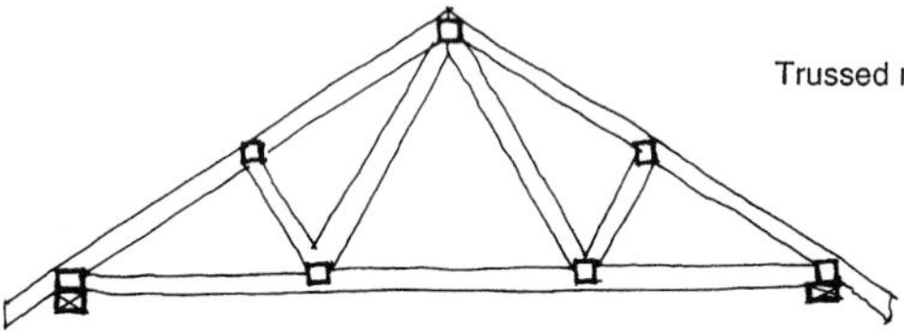

Trussed rafter

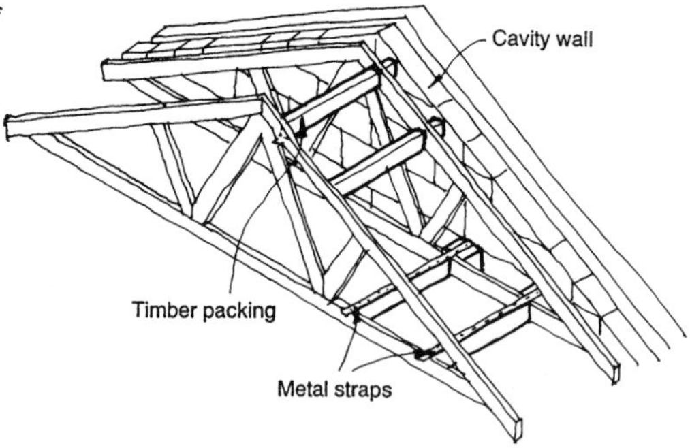

Lateral
wall support

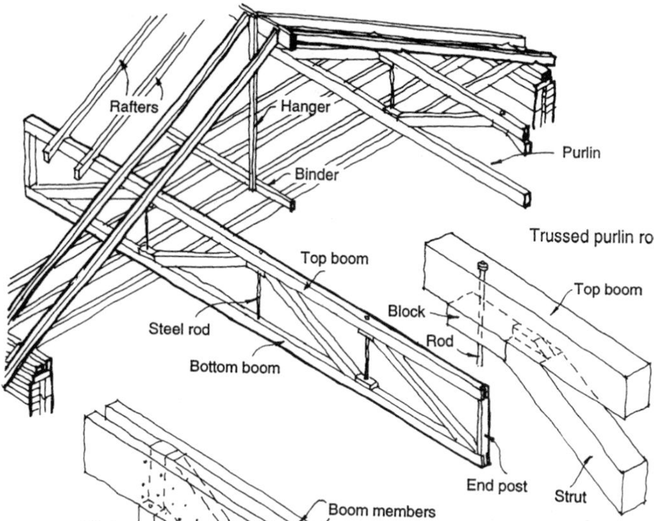

Trussed purlin roof

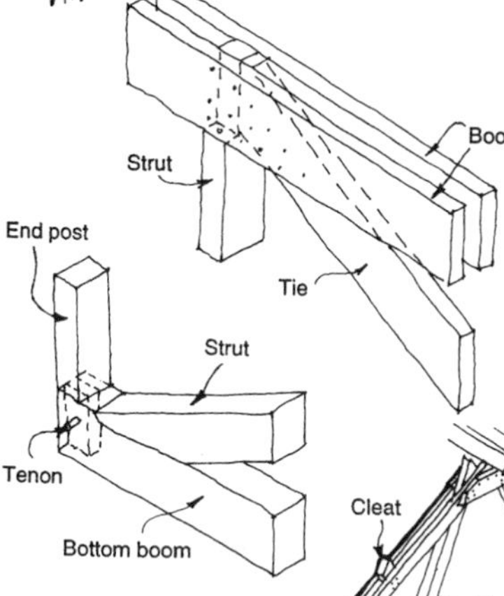

Double lap sandwich construction

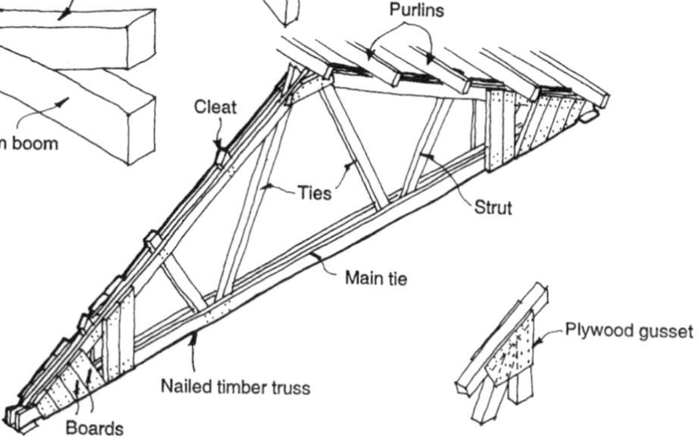

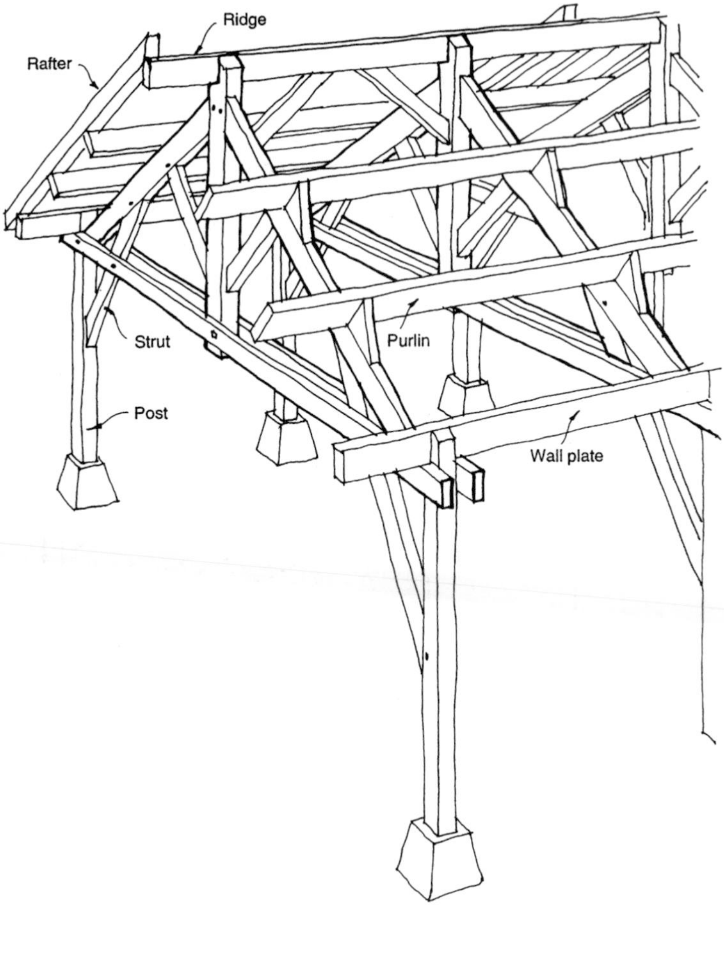
Ridge
Rafter
Strut
Purlin
Post
Wall plate

Lean-to half truss

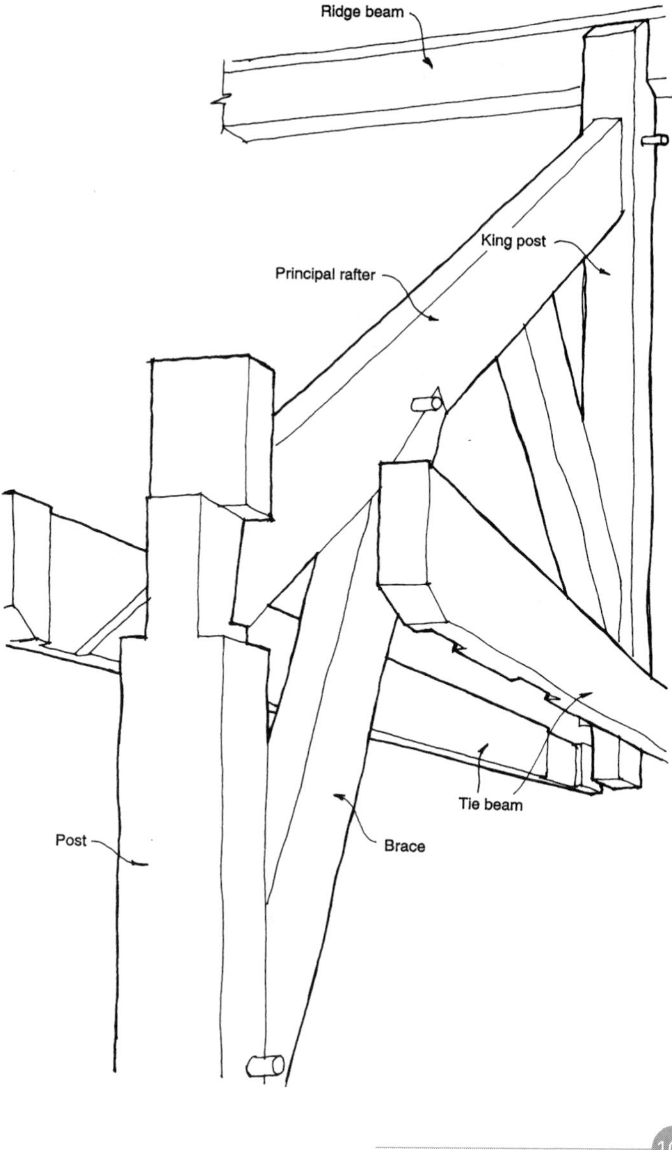
Ridge beam
King post
Principal rafter
Tie beam
Post
Brace

Details

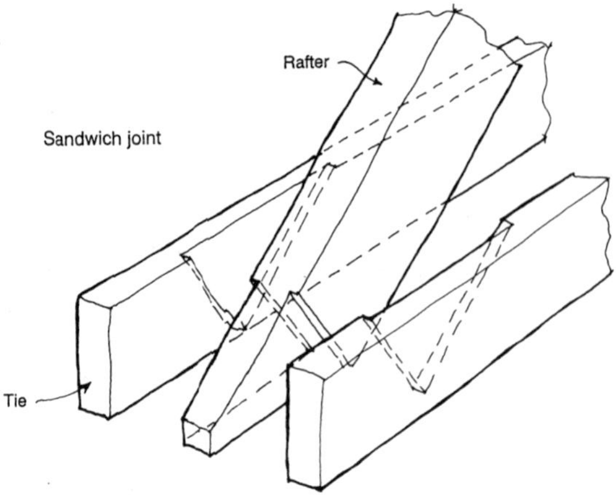
Rafter
Sandwich joint
Tie

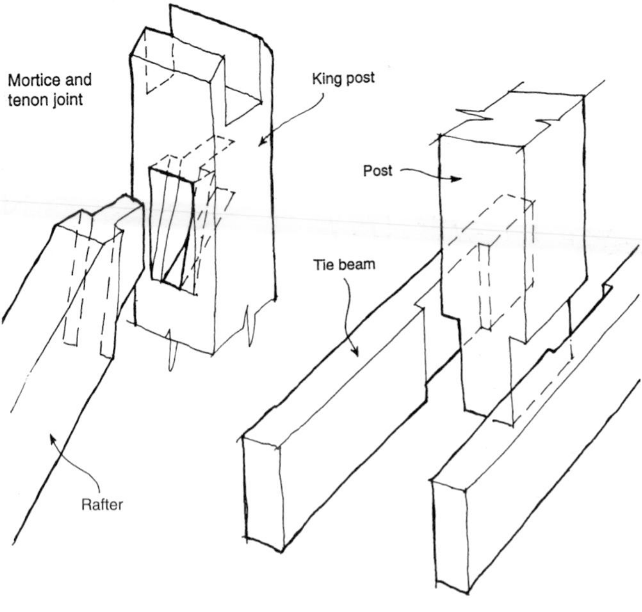
Mortice and tenon joint
King post
Post
Tie beam
Rafter

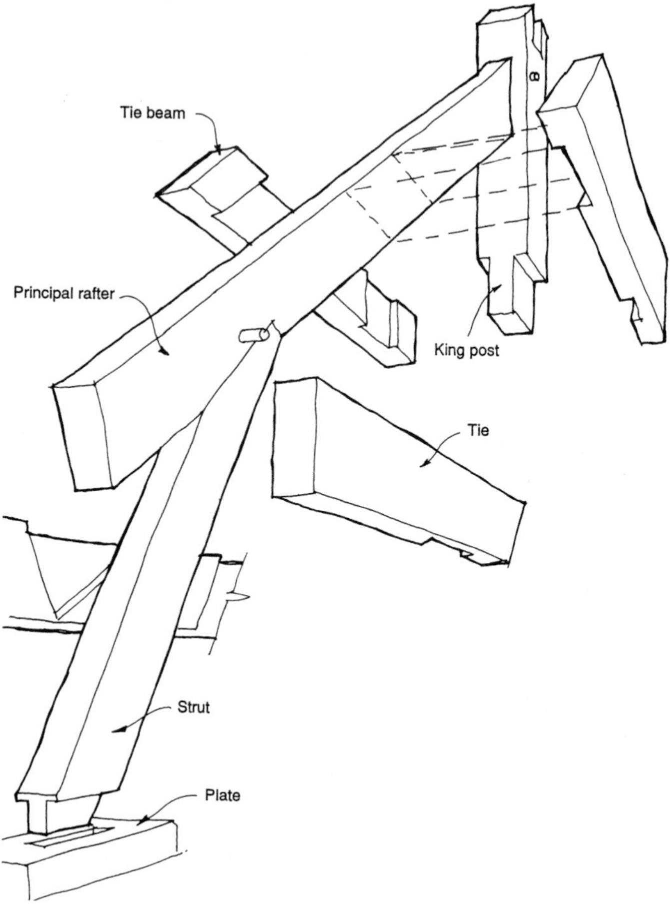
Tie beam
Principal rafter
King post
Tie
Strut
Plate

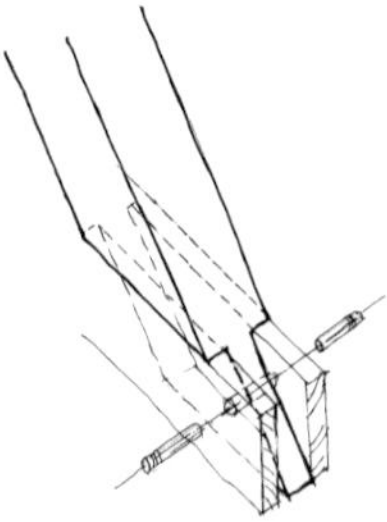
Bolt and connector sandwich joint

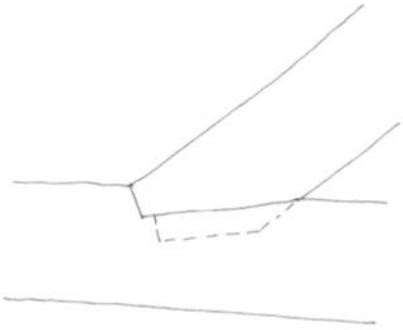
Oblique tenon

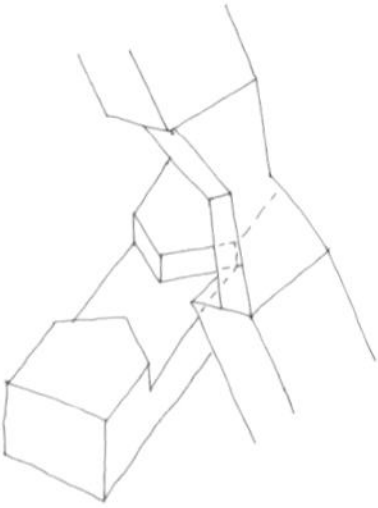
Counter – sallied cross

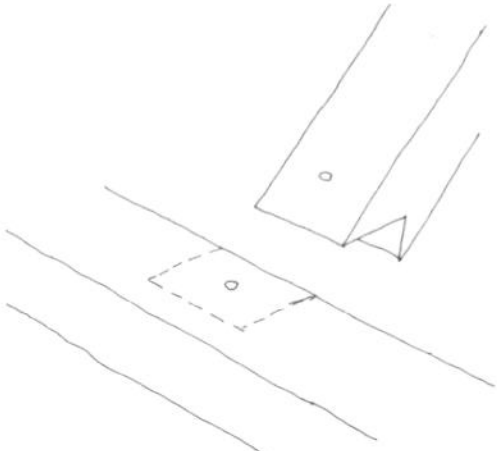
Birdsmouth joint

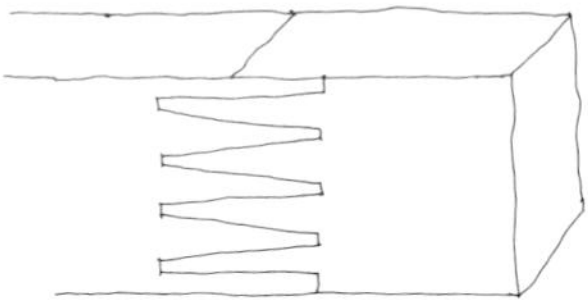
Finger joint

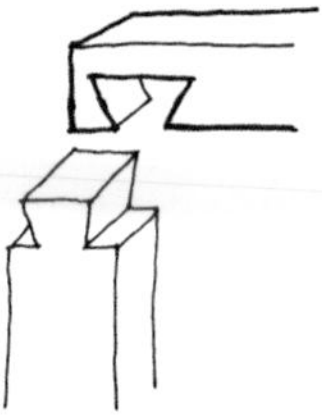
Dovetail joint

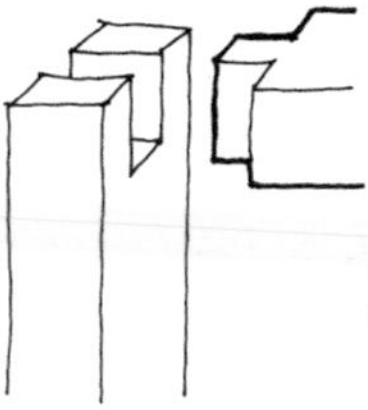
Corner locking joint

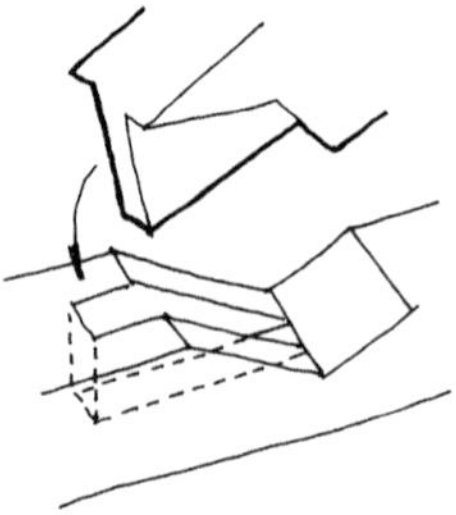
Oblique tenon

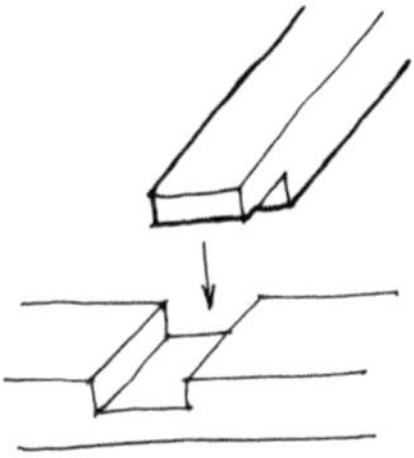
Half-lap joint

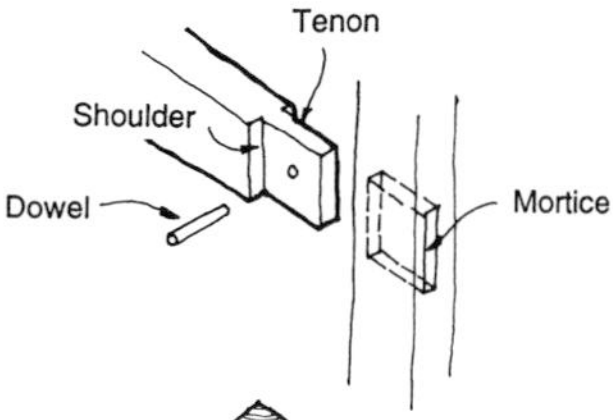

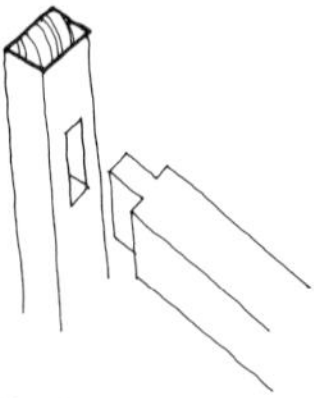
Mortice and tenon

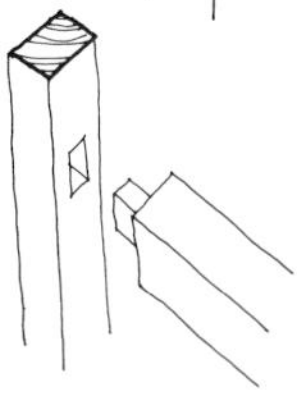
Double-shouldered tenon

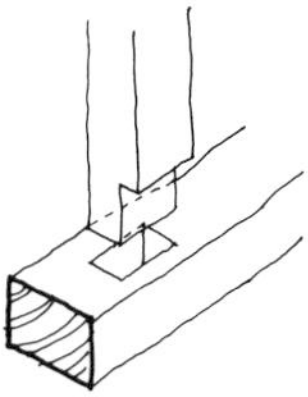
Bare-faced tenon

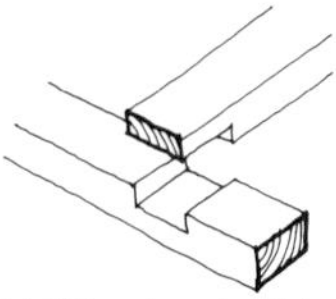
Half-face housing joint

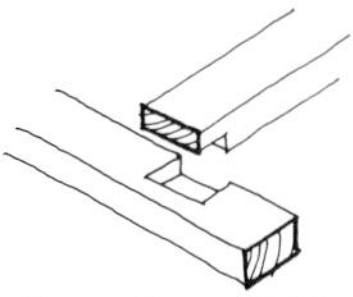
Square housed joint

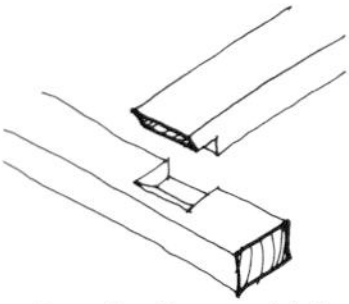
Bevelled housed joint

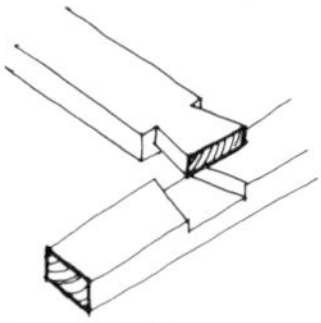
Dovetailed lap

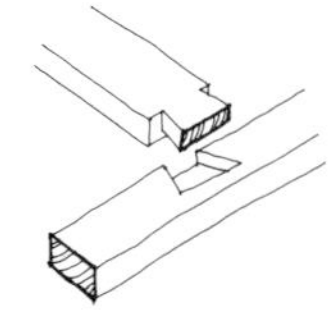
Dovetailed housing

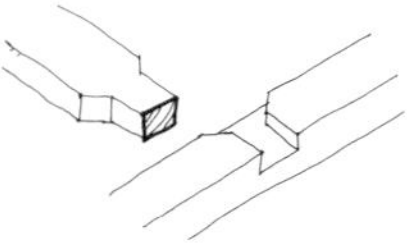
Splay-shouldered lap

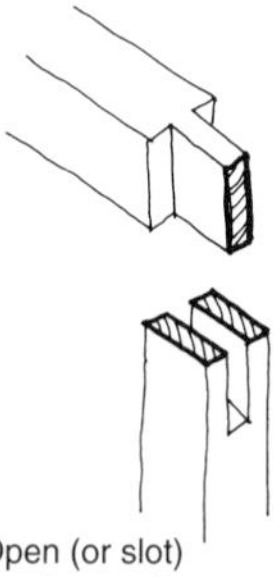
Open (or slot)
mortice and tenon

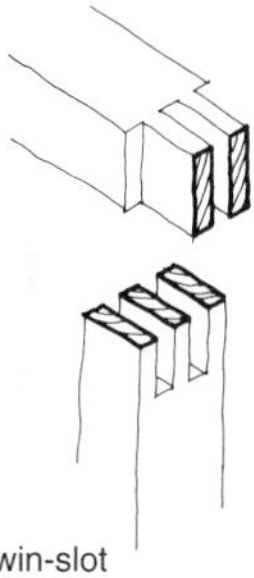
Twin-slot
mortice and tenon

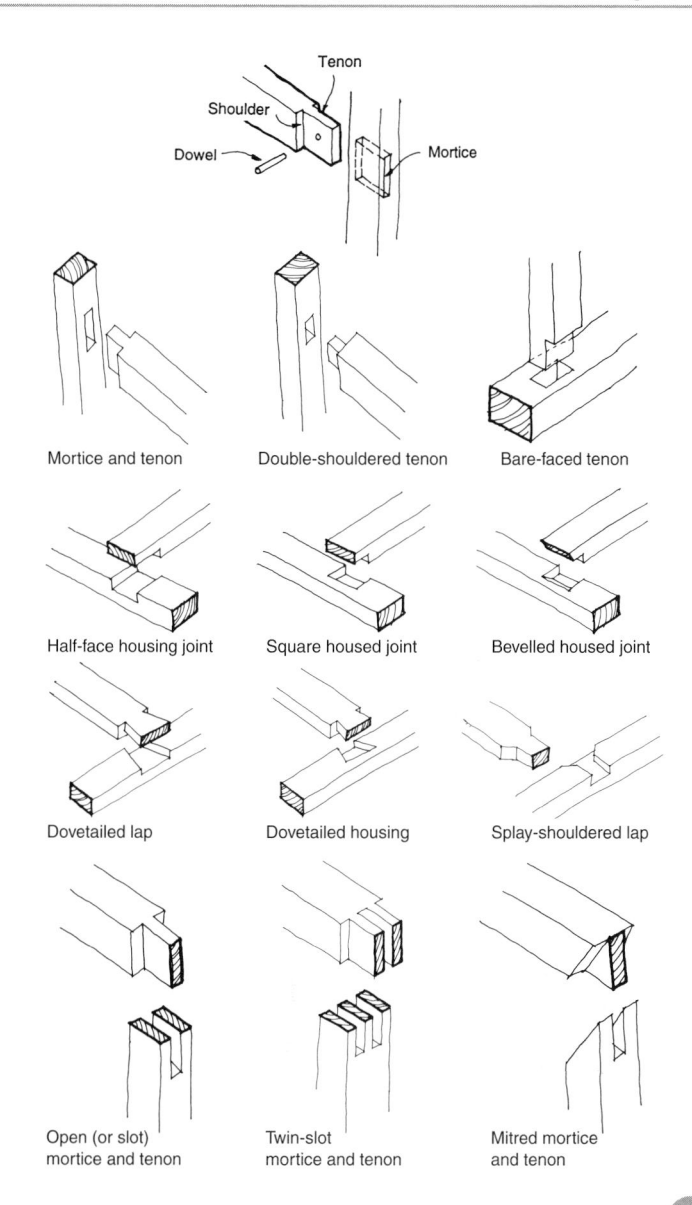

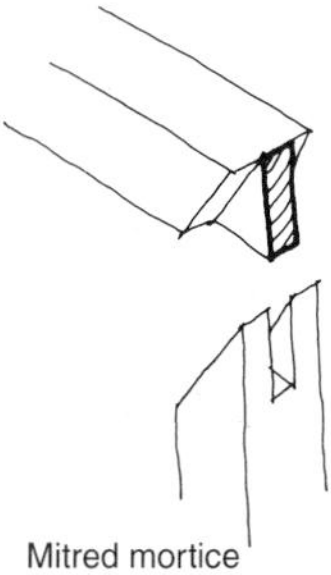
Mitred mortice
and tenon

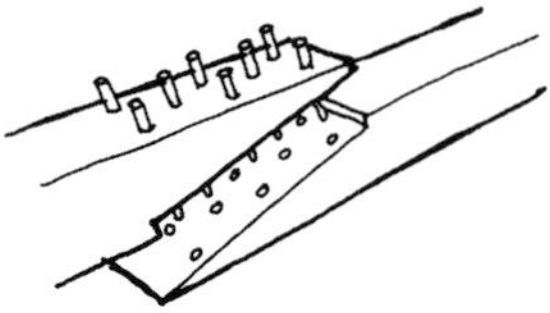

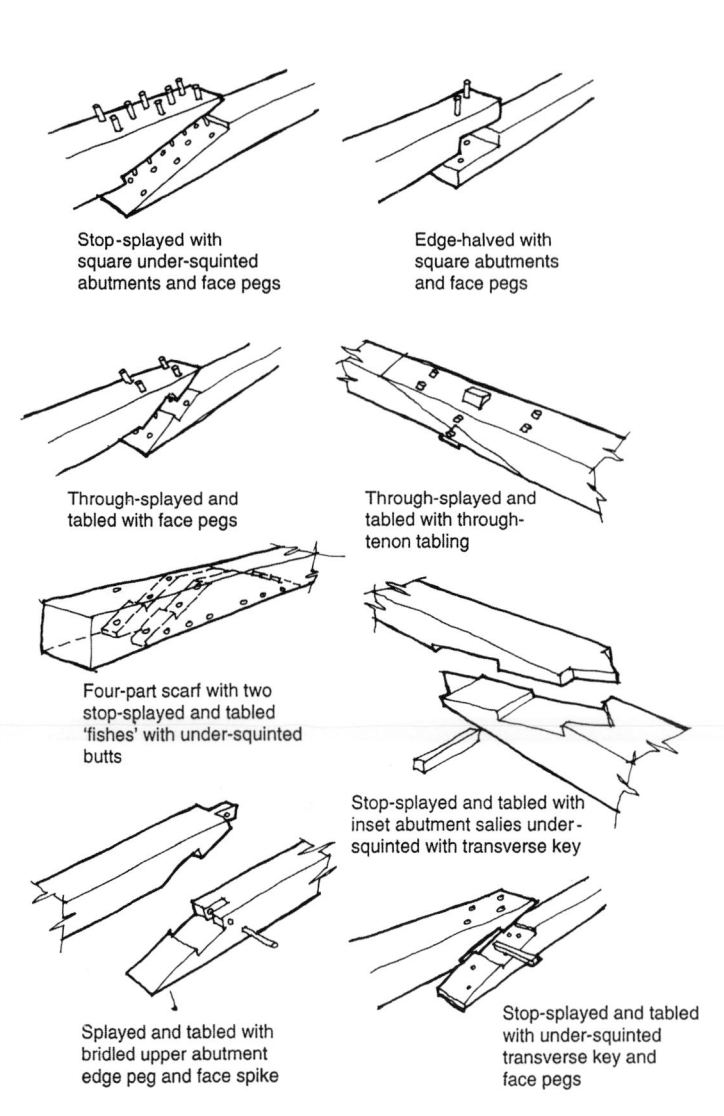

Stop-splayed with square under-squinted abutments and face pegs

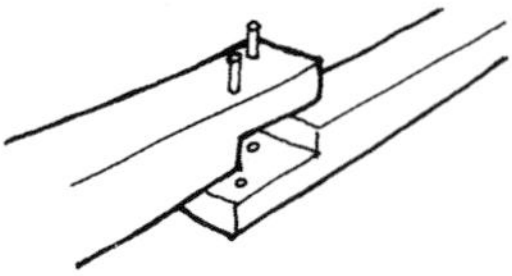

Edge-halved with square abutments and face pegs

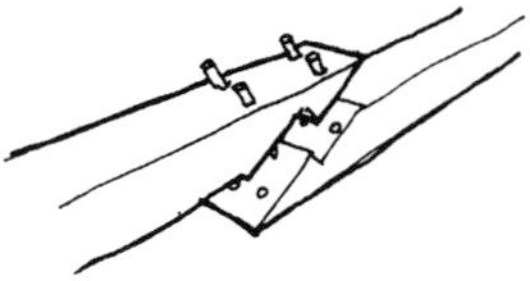

Through-splayed and tabled with face pegs

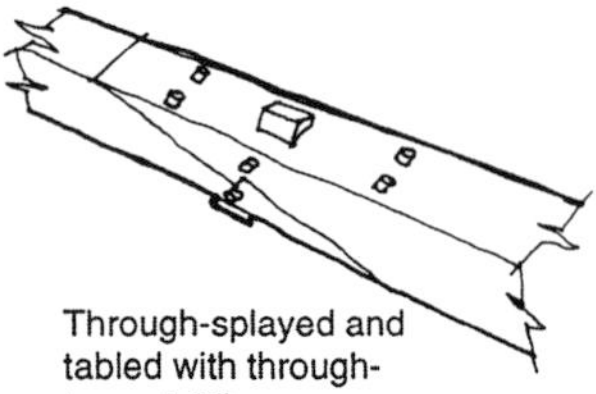

Through-splayed and tabled with through-tenon tabling

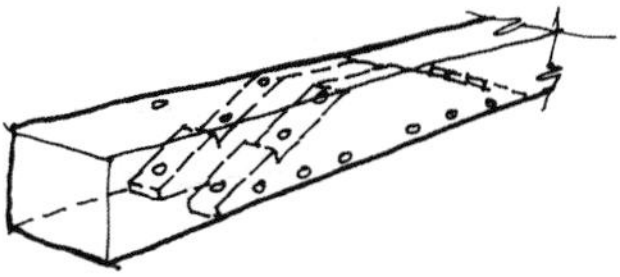

Four-part scarf with two stop-splayed and tabled 'fishes' with under-squinted butts

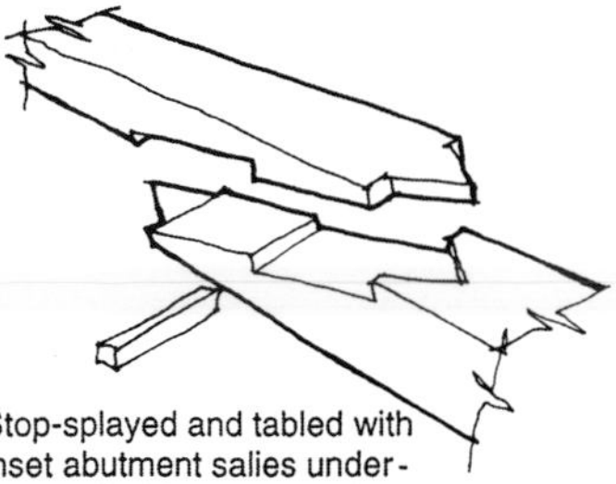

Stop-splayed and tabled with inset abutment salies under-squinted with transverse key

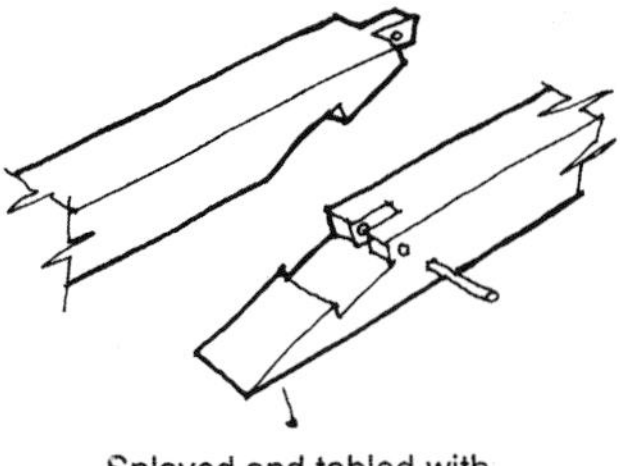

Splayed and tabled with bridled upper abutment edge peg and face spike

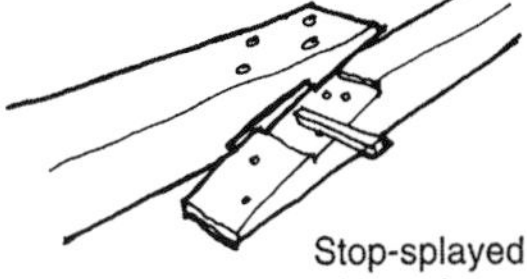

Stop-splayed and tabled with under-squinted transverse key and face pegs

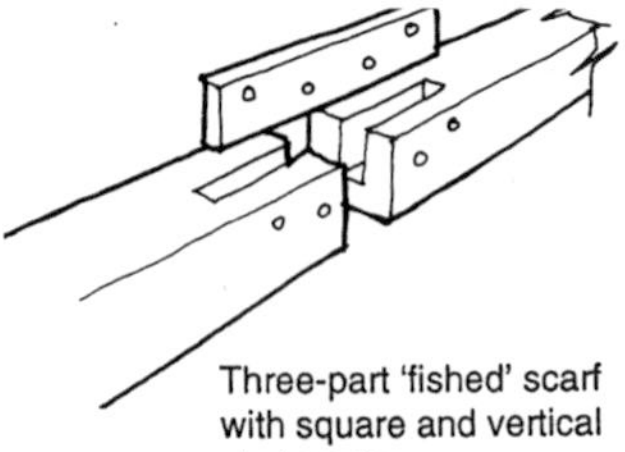

Three-part 'fished' scarf with square and vertical abutments

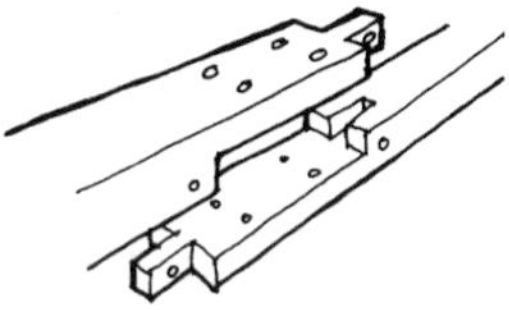

Edge-halved and stop-splayed with bridled abutments

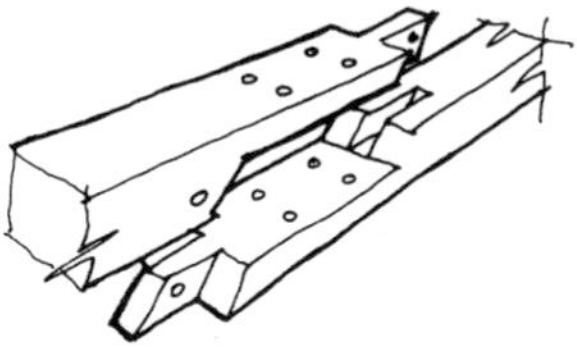

Edge-halved and bridled with over-squinted abutments

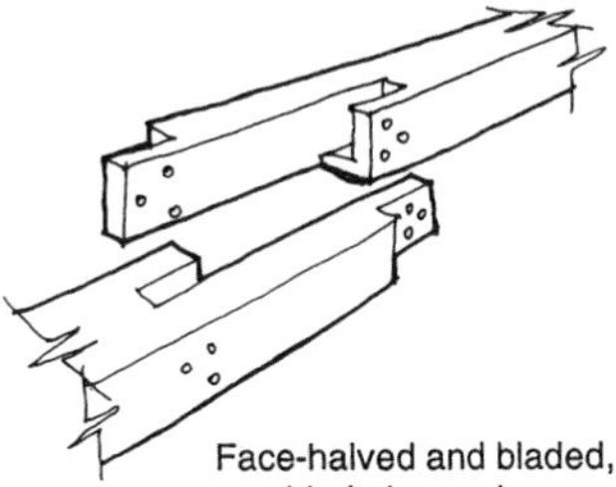

Face-halved and bladed, one blade housed, edge pegs

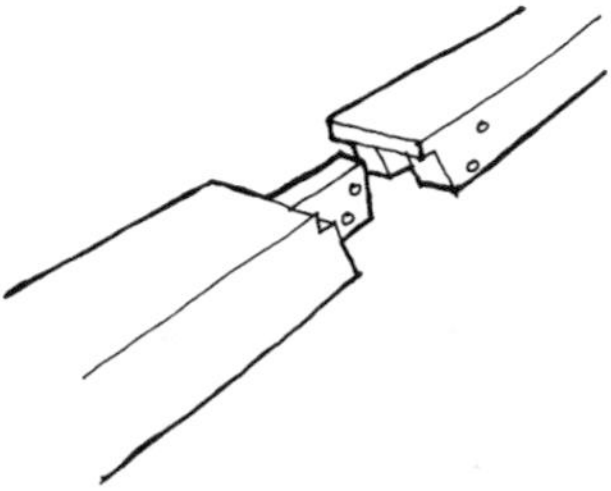

Straight bridling of three-quarter depth, with squinted abutments, overlipped face, edge pegs

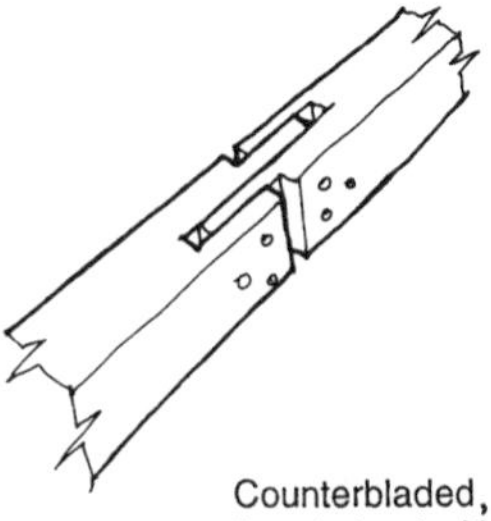

Counterbladed, face-halved with edge pegs

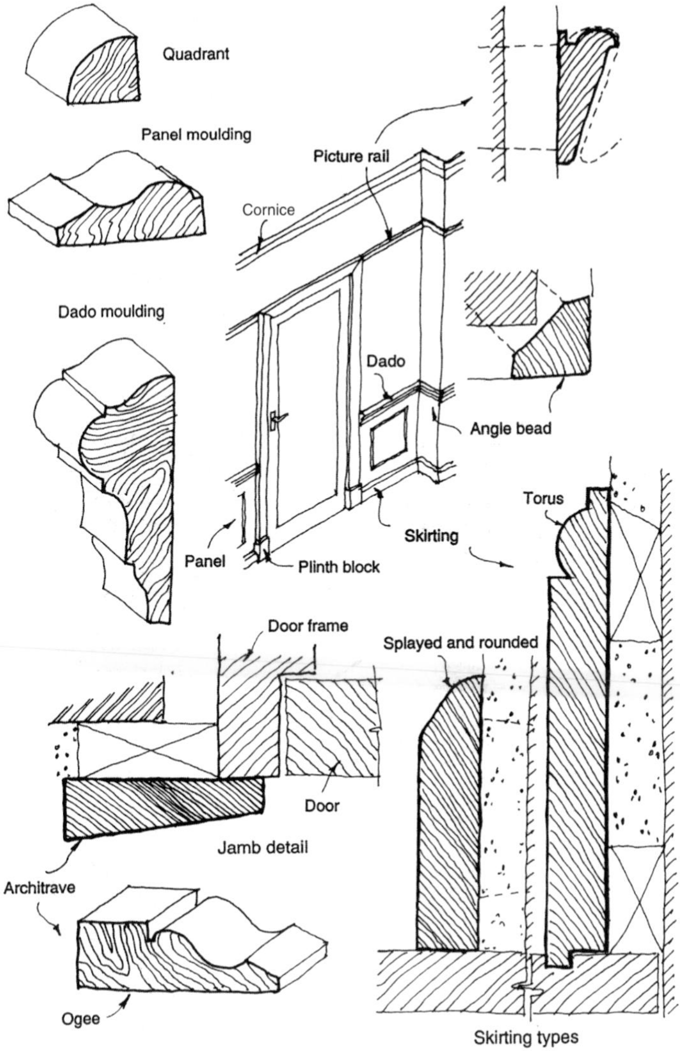
Quadrant
Panel moulding
Picture rail
Cornice
Dado moulding
Dado
Angle bead
Torus
Skirting
Panel
Plinth block
Door frame
Splayed and rounded
Door
Jamb detail
Architrave
Ogee
Skirting types

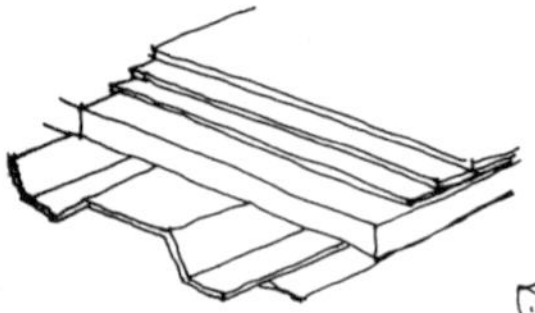

Polymer-modified polyester system (APP SBS) on metal deck

Typical mastic asphalt flat roof on concrete deck

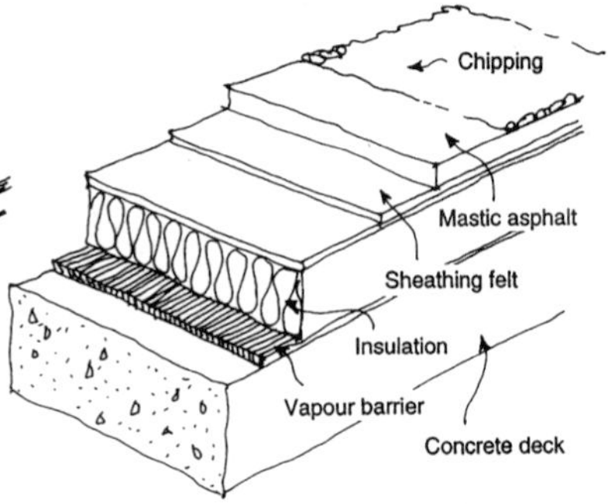

Flat roof on timber deck

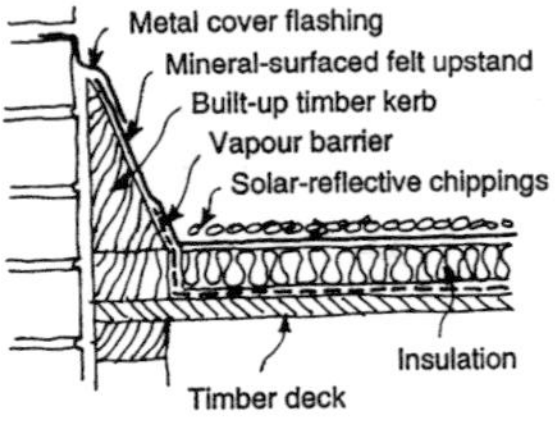

Green roof

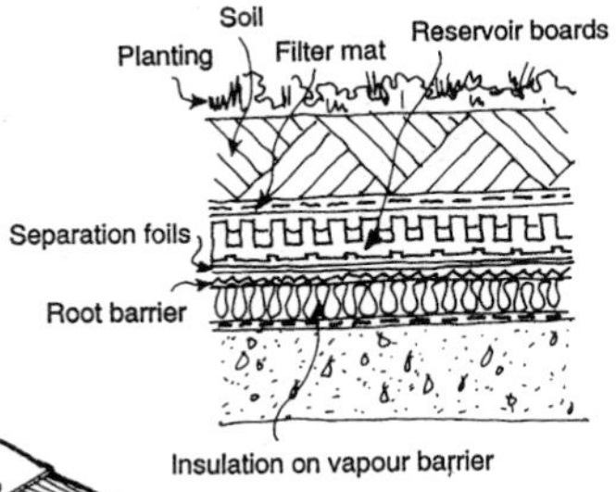

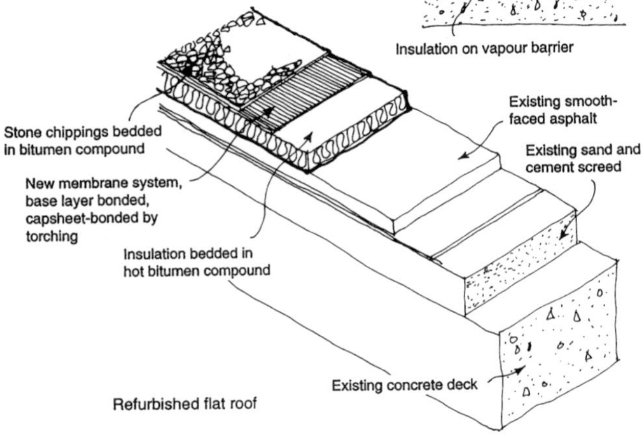

Refurbished flat roof

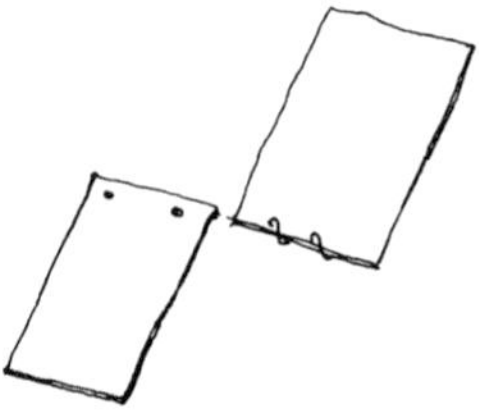

Slate

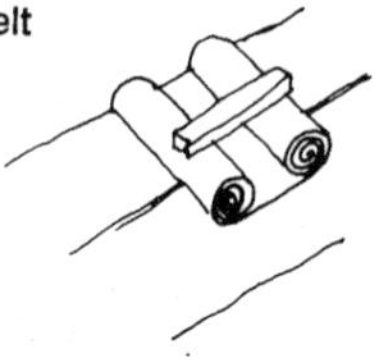

Felt

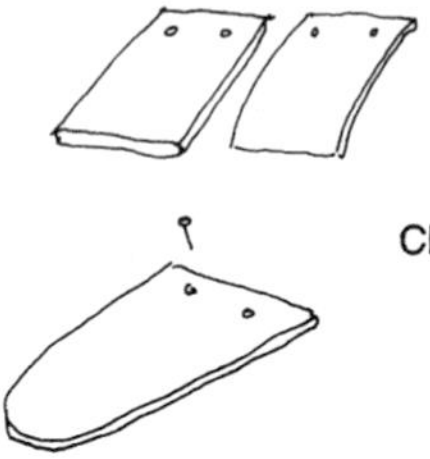

Clay plain tiles

Roman tiles

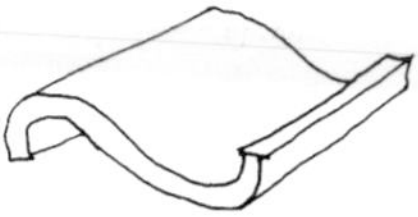

Pantile

Stone tile

Thatch

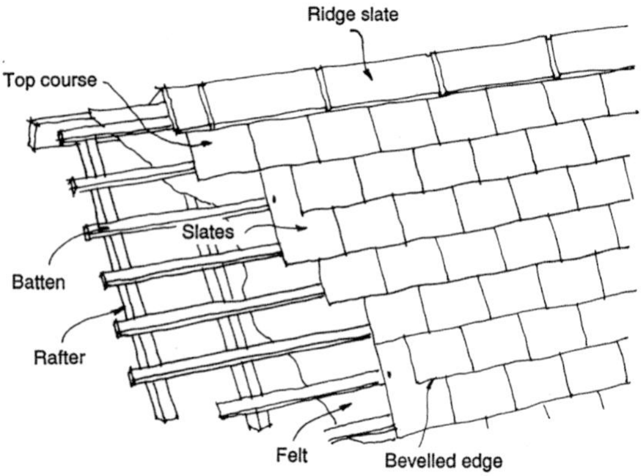
Ridge slate
Top course
Slates
Batten
Rafter
Felt
Bevelled edge

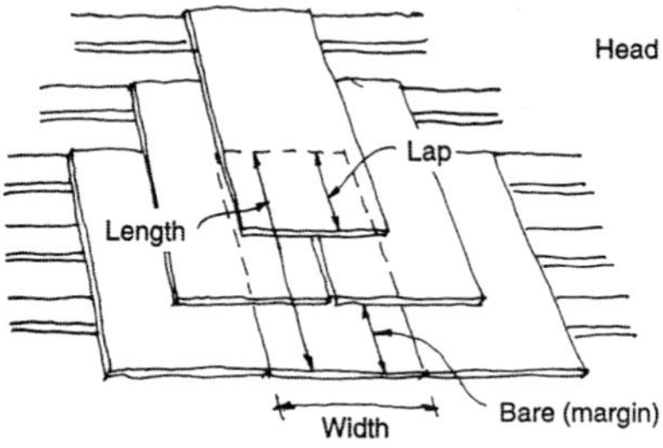
Lap
Length
Width
Bare (margin)

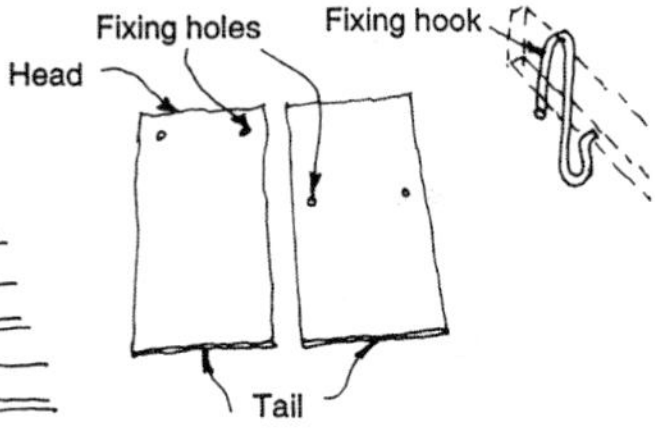
Fixing holes
Fixing hook
Head
Tail

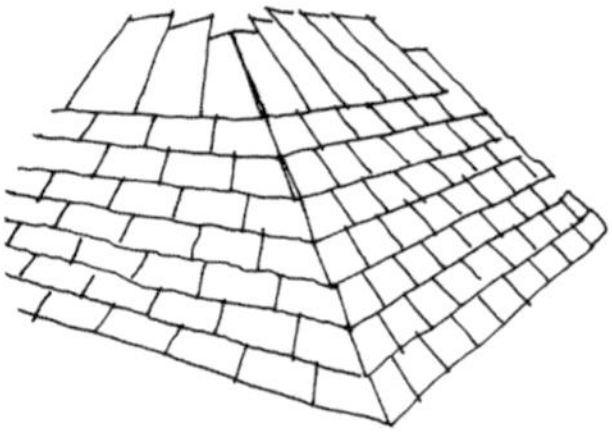
Hip with cut and mitred slates

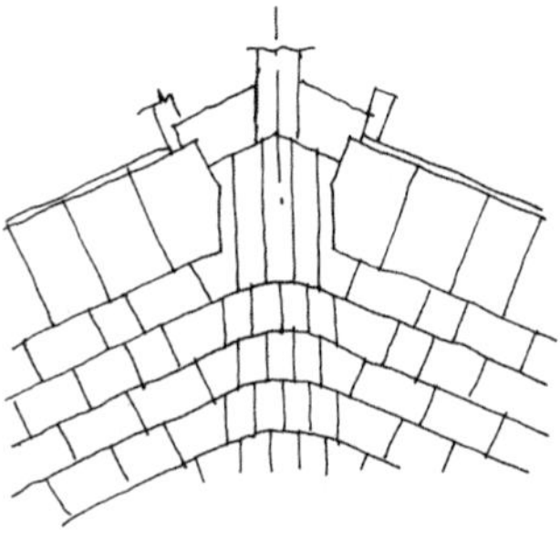
Rounded valley

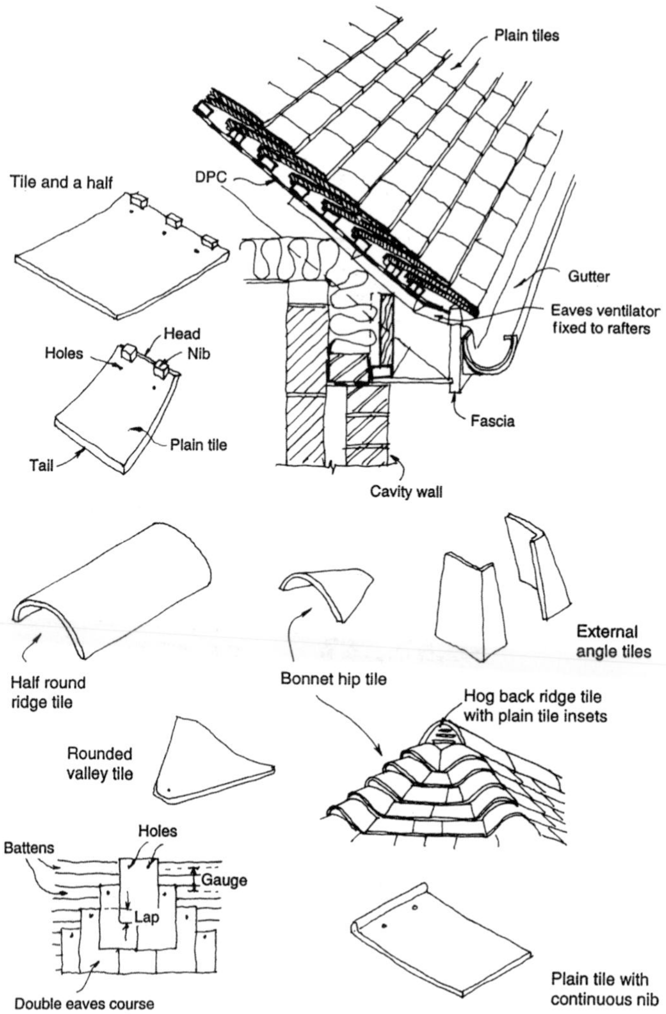
Plain tiles
Tile and a half
DPC
Gutter
Eaves ventilator
fixed to rafters
Head
Holes
Nib
Plain tile
Tail
Fascia
Cavity wall
External
angle tiles
Half round
ridge tile
Bonnet hip tile
Hog back ridge tile
with plain tile insets
Rounded
valley tile
Holes
Battens
Gauge
Lap
Double eaves course
Plain tile with
continuous nib

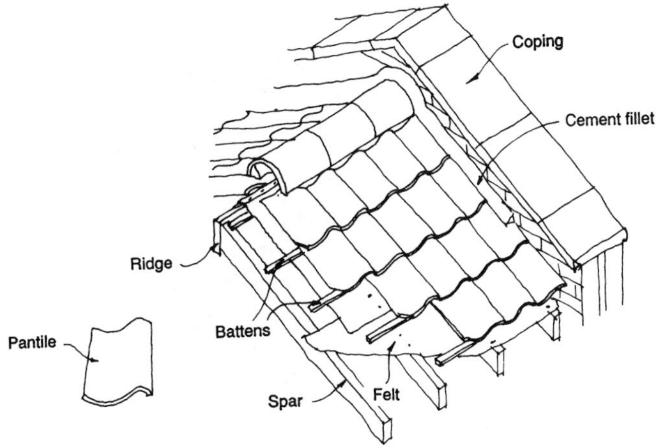
Coping
Cement fillet
Ridge
Pantile
Battens
Spar
Felt

Roman (or Italian) tiling

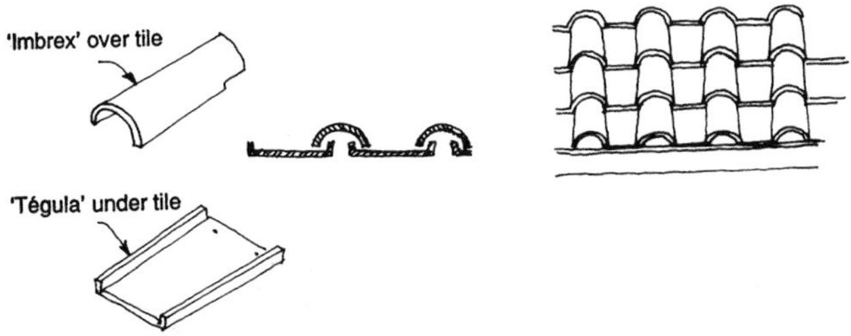
'Imbrex' over tile
'Tégula' under tile

Spanish tiling

Over
Head
Under

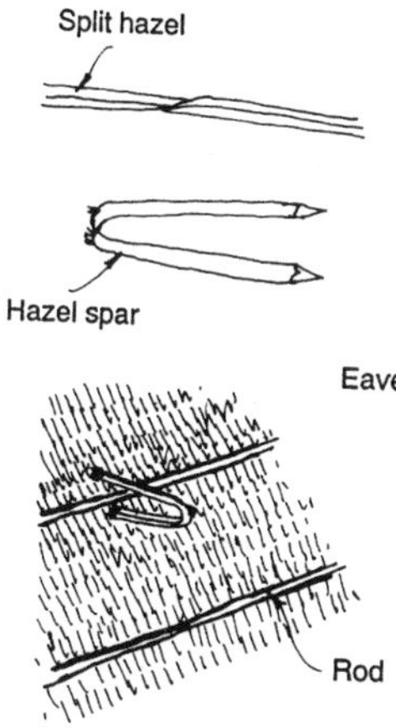
Split hazel
Hazel spar
Rod

Spar fixing

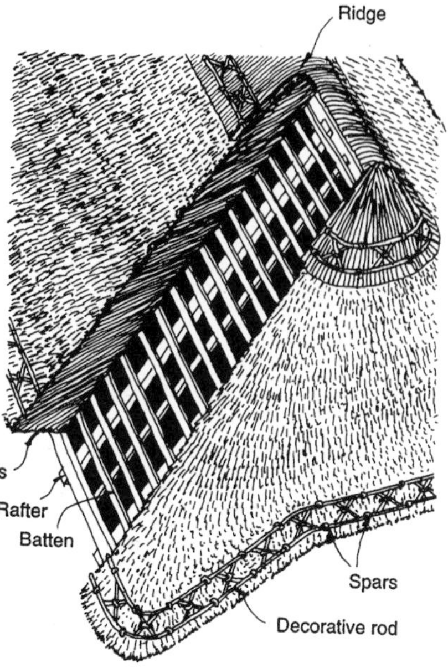
Ridge
Eaves
Rafter
Batten
Spars
Decorative rod

Ridge details

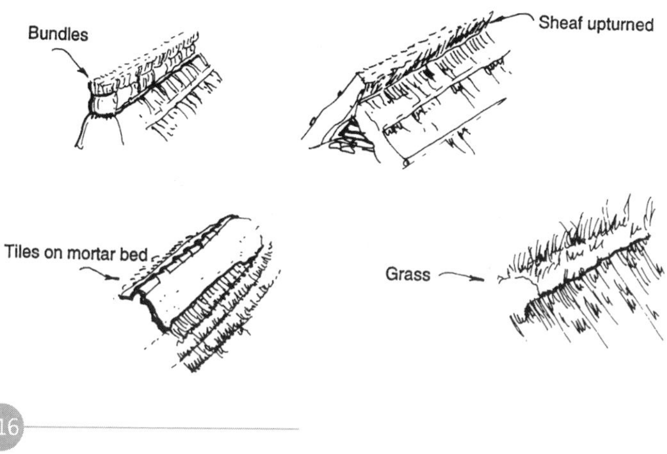
Bundles
Sheaf upturned
Tiles on mortar bed
Grass

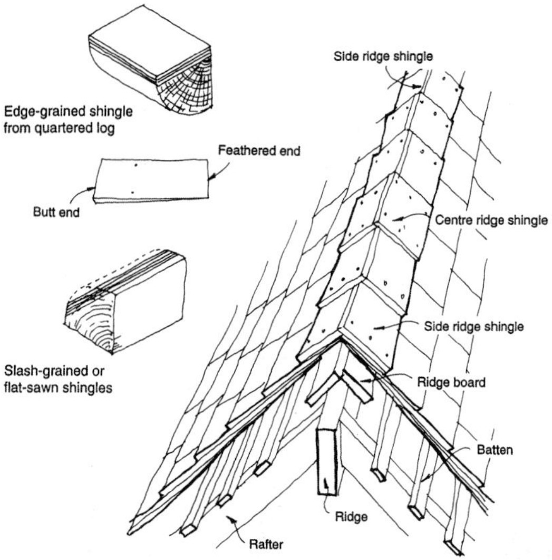
Edge-grained shingle
from quartered log
Feathered end
Butt end
Slash-grained or
flat-sawn shingles
Side ridge shingle
Centre ridge shingle
Side ridge shingle
Ridge board
Batten
Ridge
Rafter

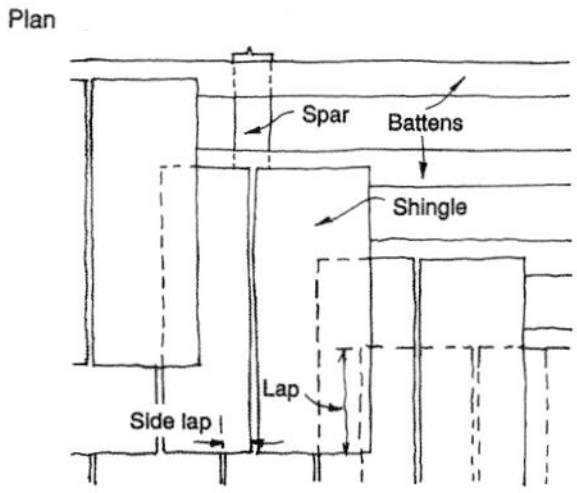
Plan
Spar
Battens
Shingle
Lap
Side lap

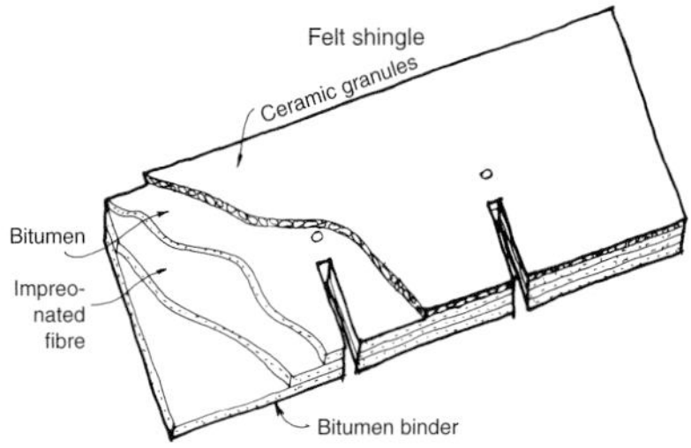
Felt shingle
Ceramic granules
Bitumen
Impreo-
nated
fibre
Bitumen binder

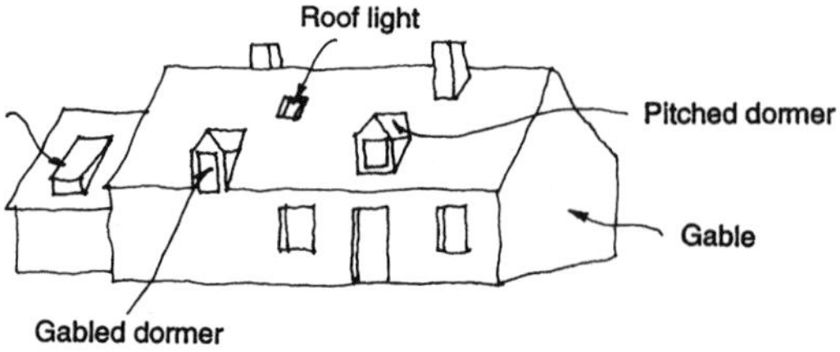
Roof light
Pitched dormer
Gable
Gabled dormer

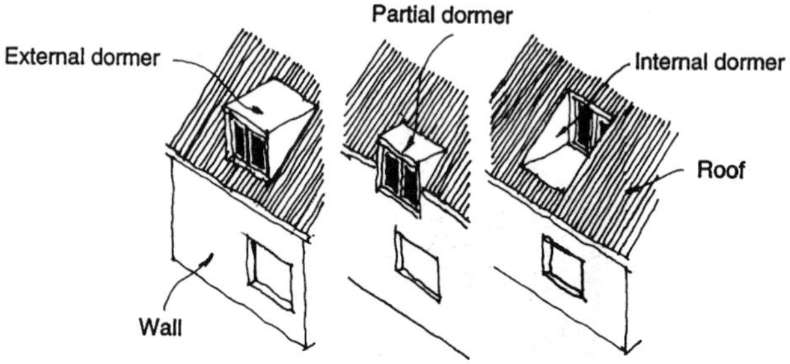
Partial dormer
External dormer
Internal dormer
Roof
Wall

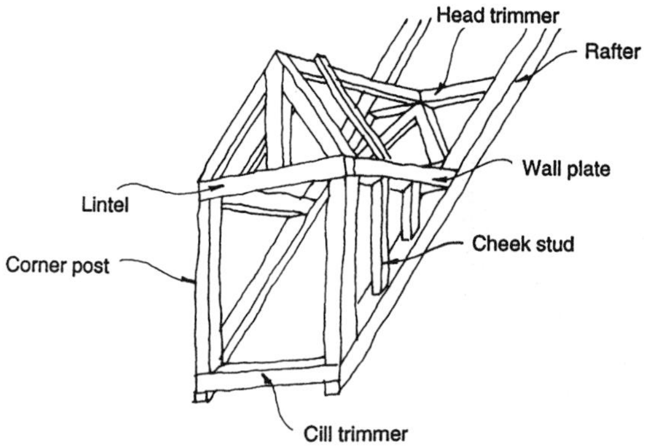
Head trimmer
Rafter
Wall plate
Lintel
Cheek stud
Corner post
Cill trimmer

Polygonal piended

Lead roofed rectangular

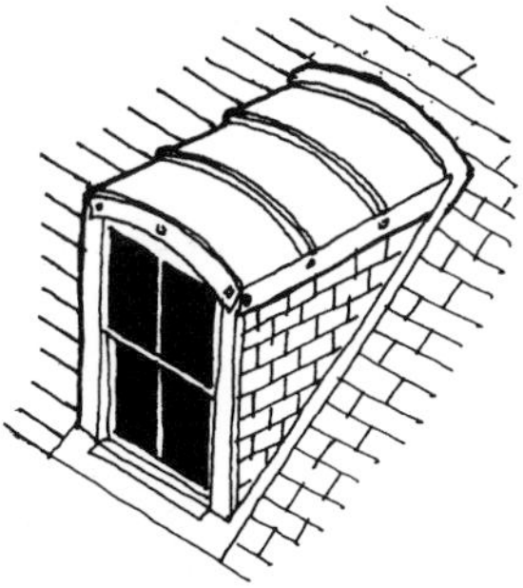

Rectangular

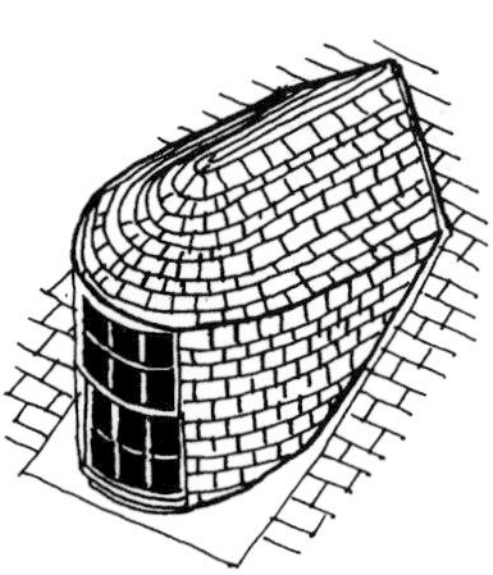

Bowed front

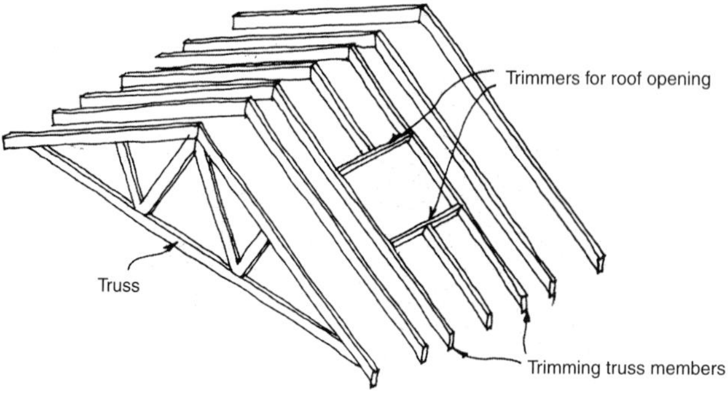
Trimmers for roof opening
Truss
Trimming truss members

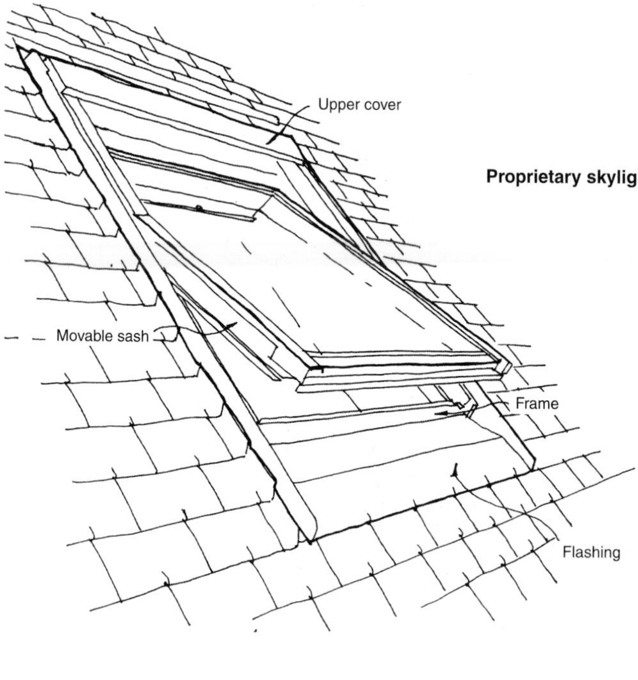
Upper cover
Movable sash
Frame
Flashing

Proprietary skylight

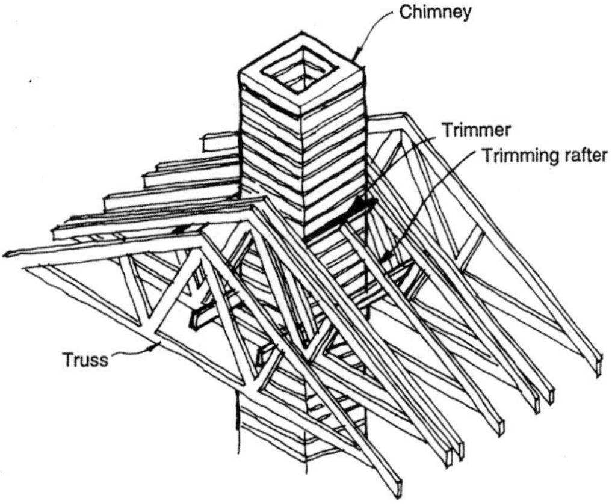
Chimney
Trimmer
Trimming rafter
Truss

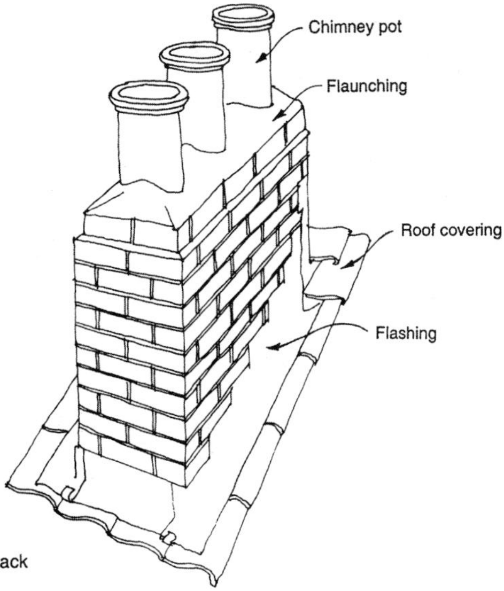
Chimney pot
Flaunching
Roof covering
Flashing

Chimney stack

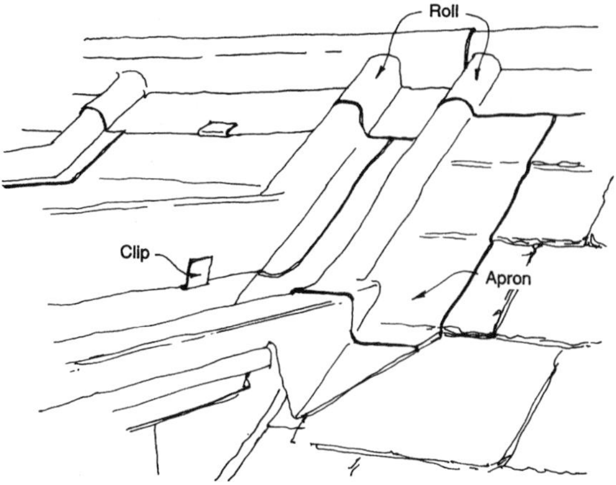

Bossed wood-cored roll

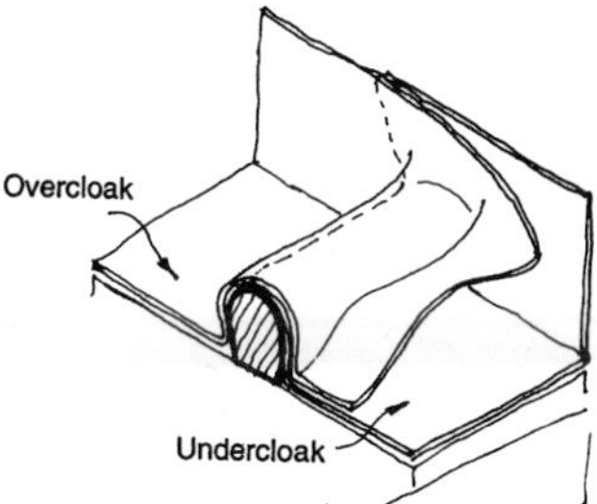

Leadburned wood-cored roll.

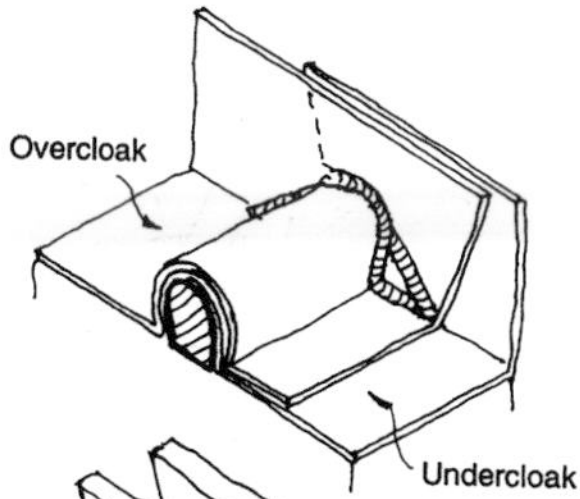

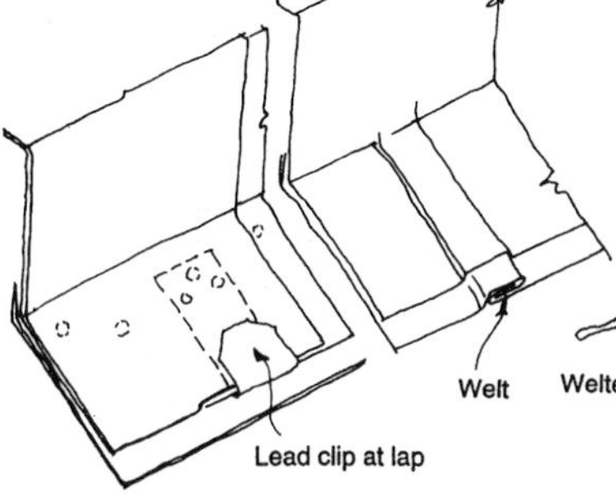

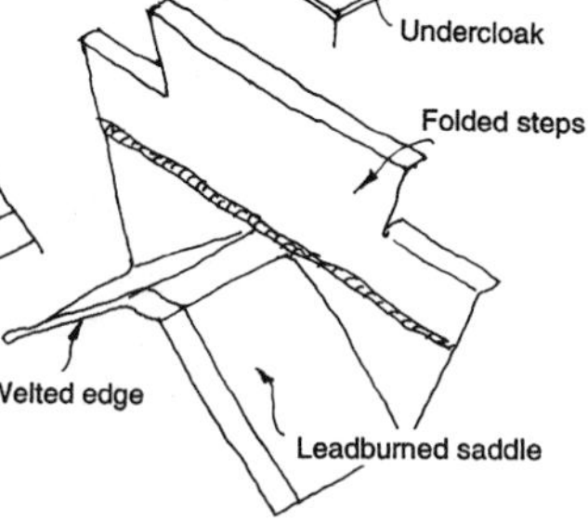

Junction roof/wall

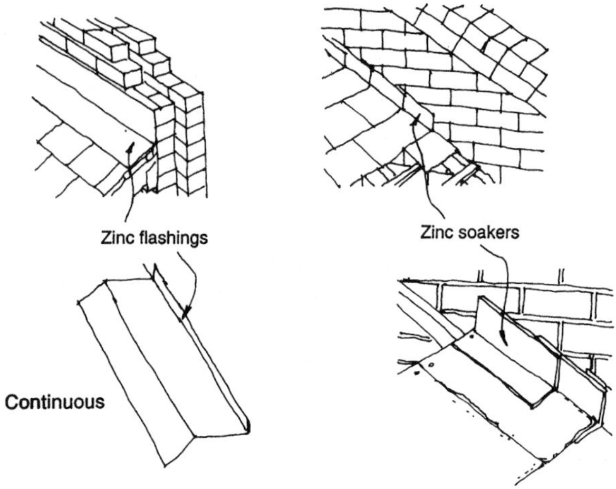

Valley

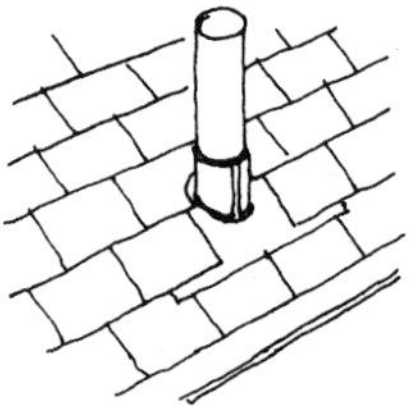

Pipe through roof

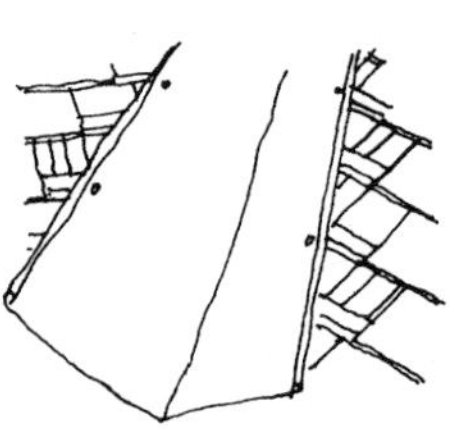

Valley liner zinc sheet

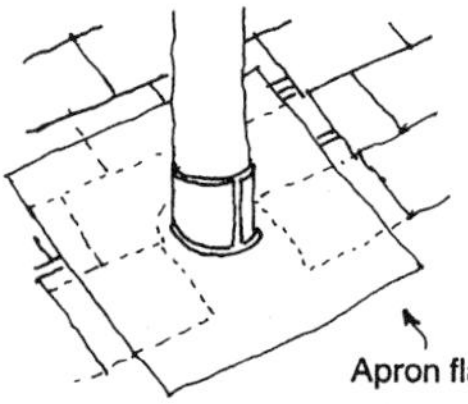

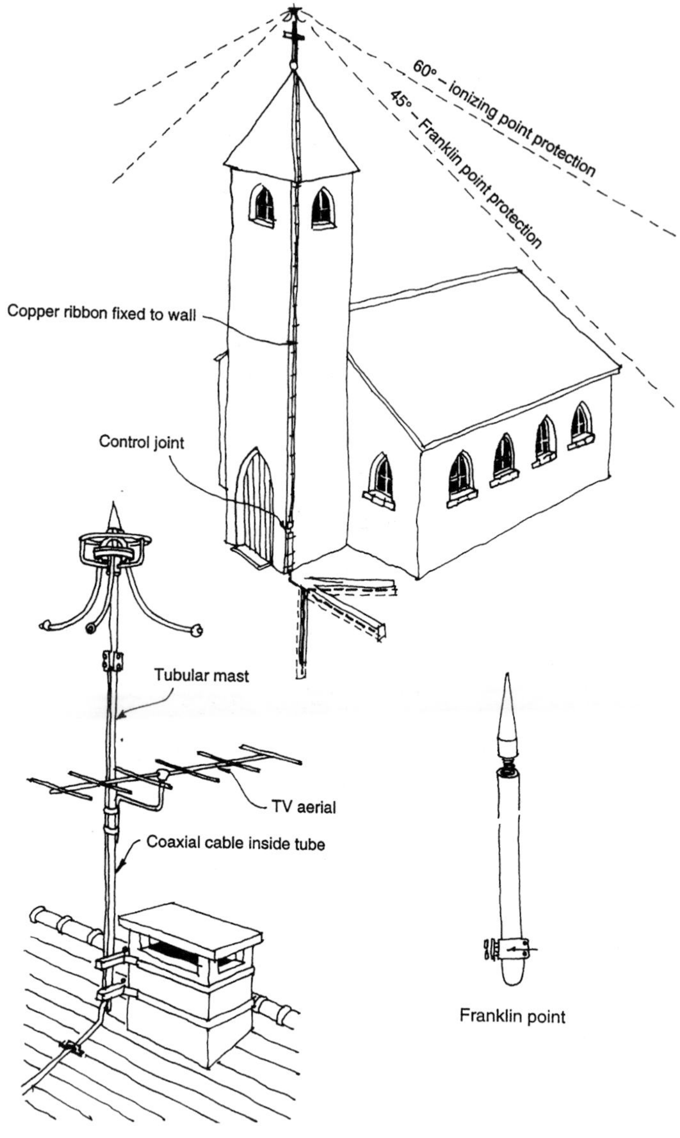
60° – ionizing point protection
45° – Franklin point protection
Copper ribbon fixed to wall
Control joint
Tubular mast
TV aerial
Coaxial cable inside tube
Franklin point

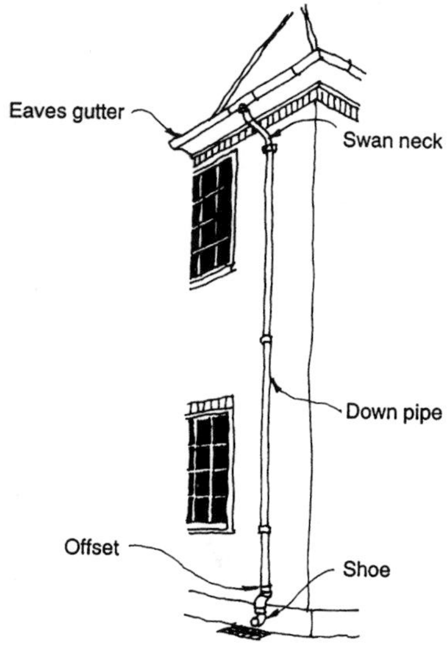
Eaves gutter
Swan neck
Down pipe
Offset
Shoe

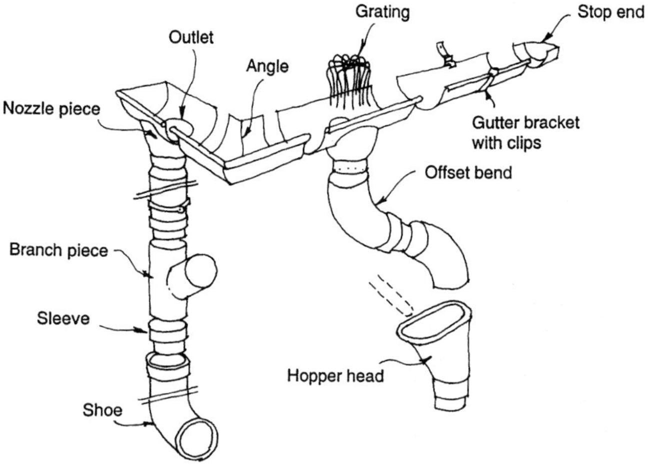
Grating
Stop end
Outlet
Angle
Nozzle piece
Gutter bracket
with clips
Offset bend
Branch piece
Sleeve
Hopper head
Shoe

Types of staircases

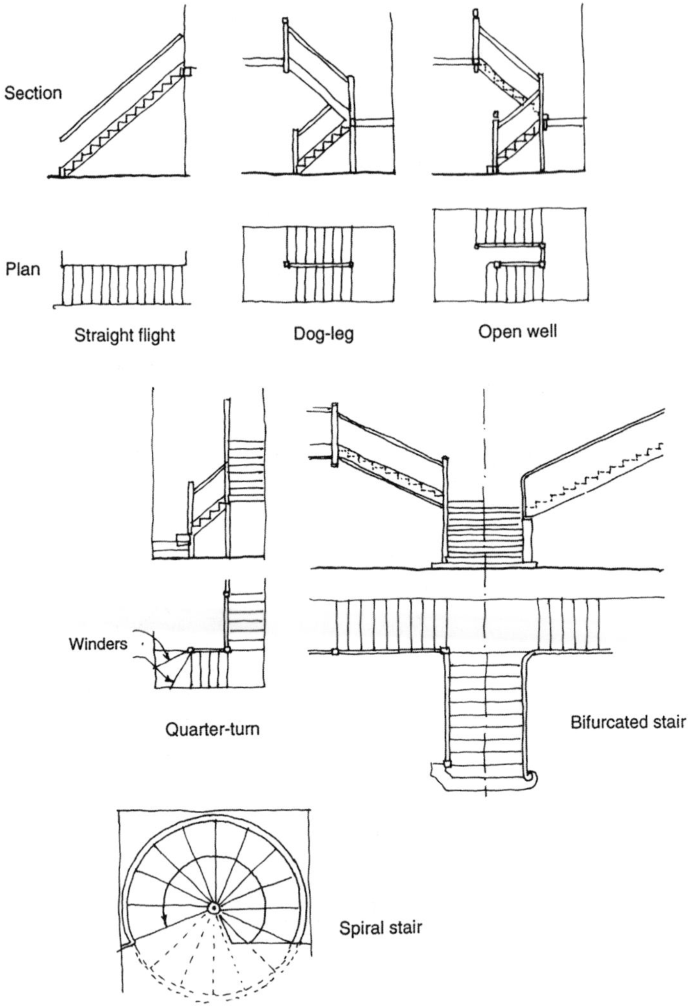
Section
Plan
Straight flight
Dog-leg
Open well
Winders
Quarter-turn
Bifurcated stair
Spiral stair

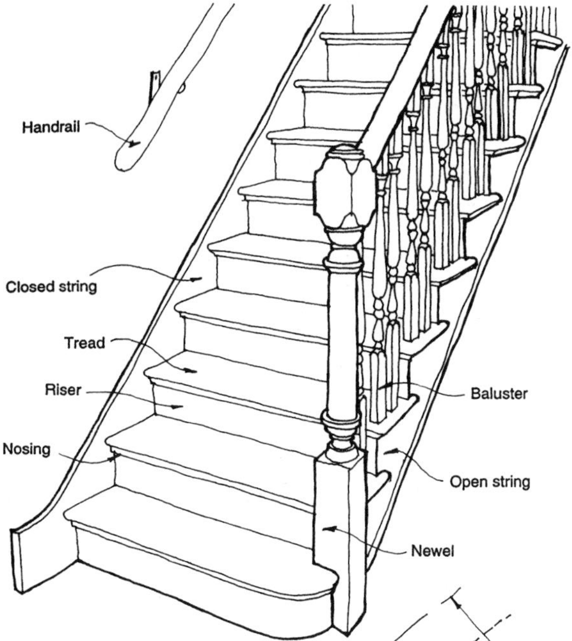
Handrail
Closed string
Tread
Riser
Nosing
Baluster
Open string
Newel

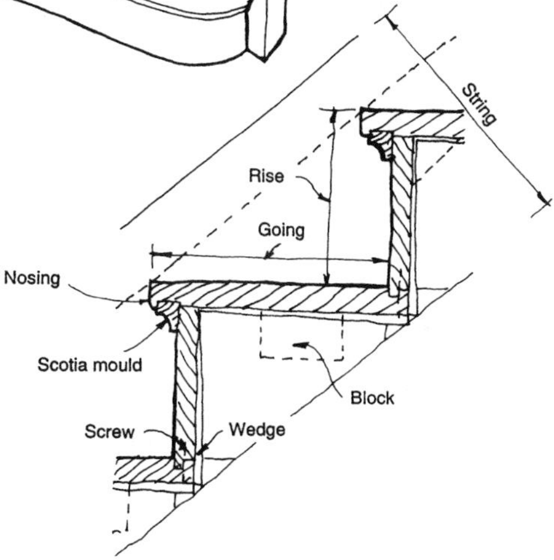
String
Rise
Going
Nosing
Scotia mould
Block
Screw
Wedge

Samba stair
(staggered steps)

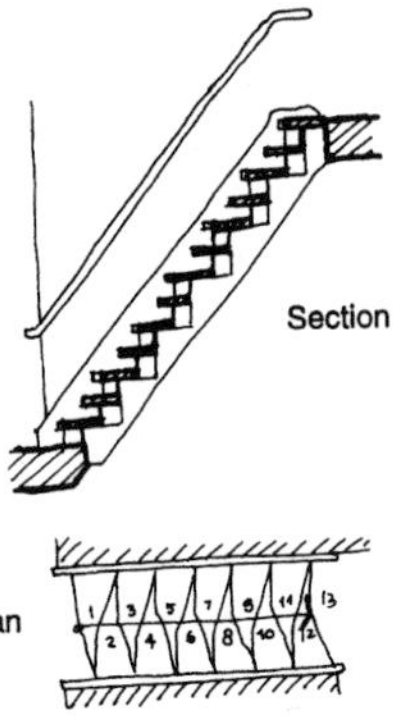

Section

Plan

Concealed
cat ladder

Retractable
loft ladder

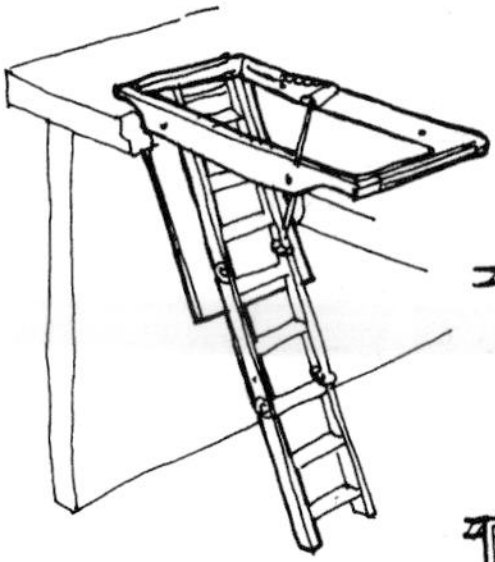

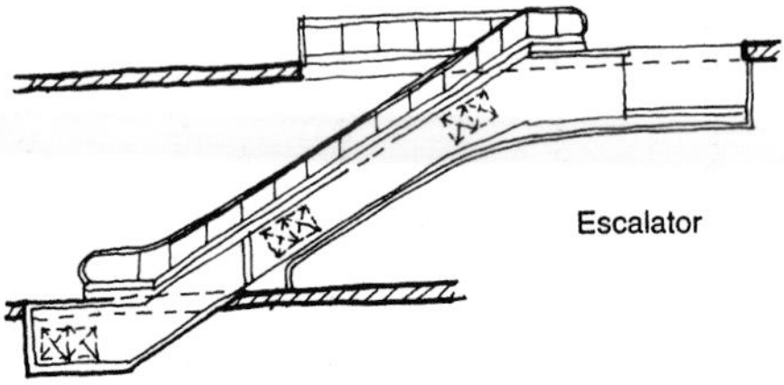

Escalator

Escalator types

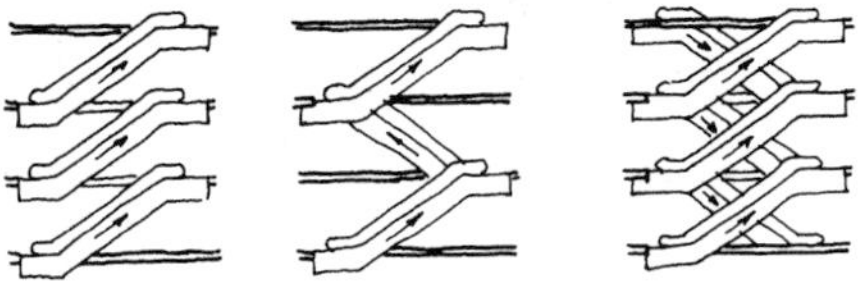

Superimposed Crossover Double-crossover

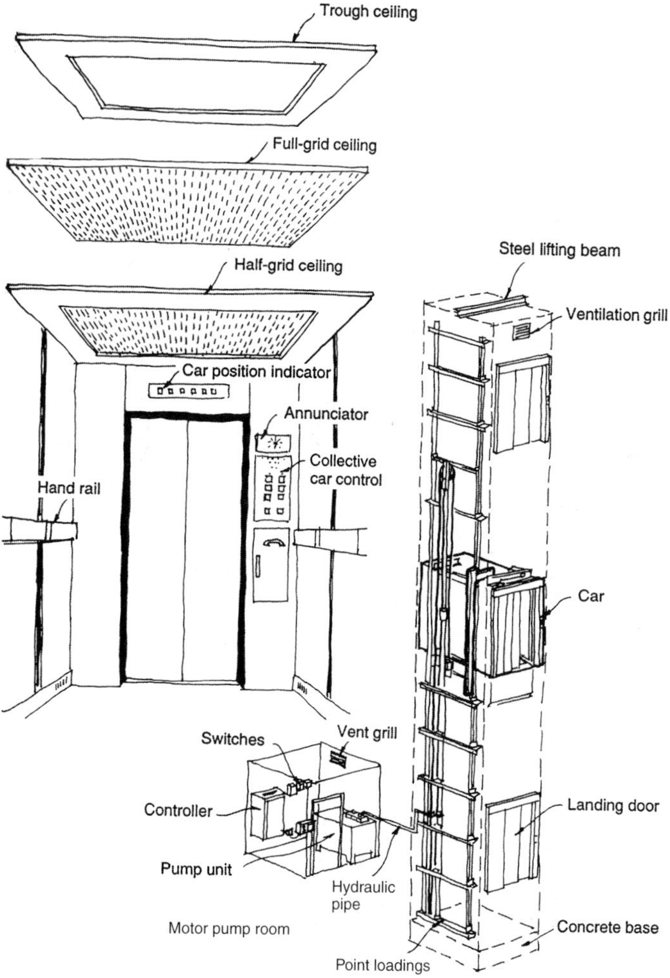

Hydraulic lift well

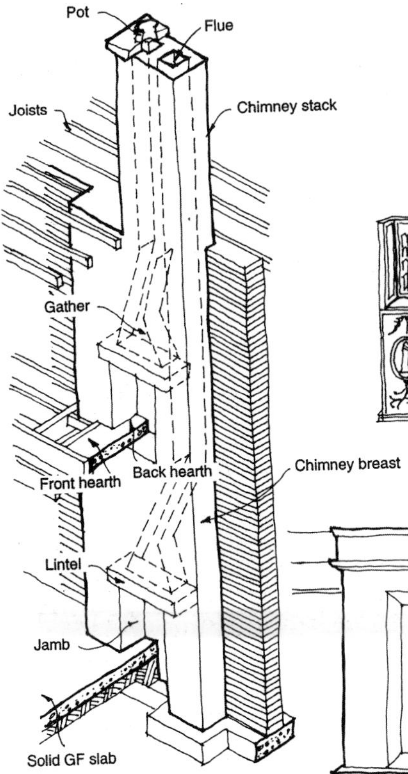
Pot
Flue
Joists
Chimney stack
Gather
Front hearth
Back hearth
Chimney breast
Lintel
Jamb
Solid GF slab

Hob grate

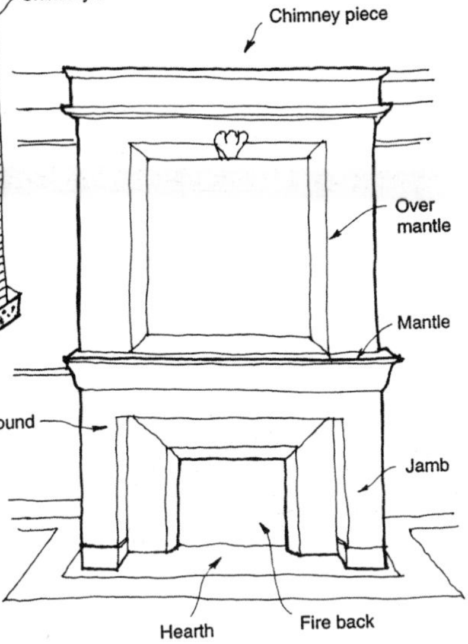
Chimney piece
Over mantle
Mantle
Fire surround
Jamb
Hearth
Fire back

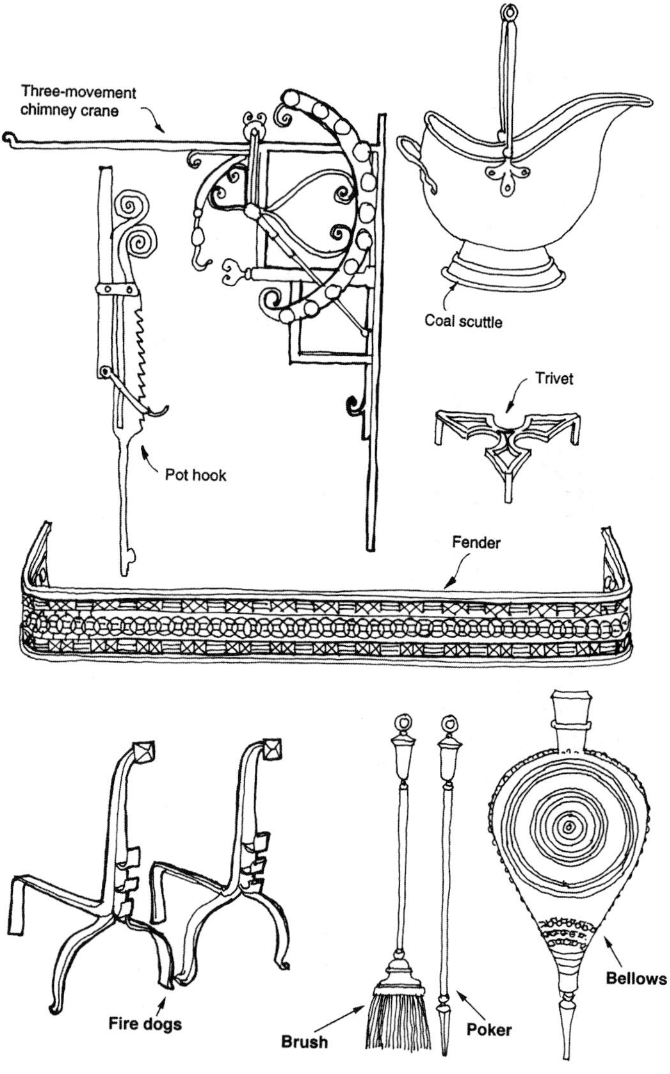
Three-movement
chimney crane
Coal scuttle
Trivet
Pot hook
Fender
Fire dogs
Brush
Poker
Bellows

Timber floor construction

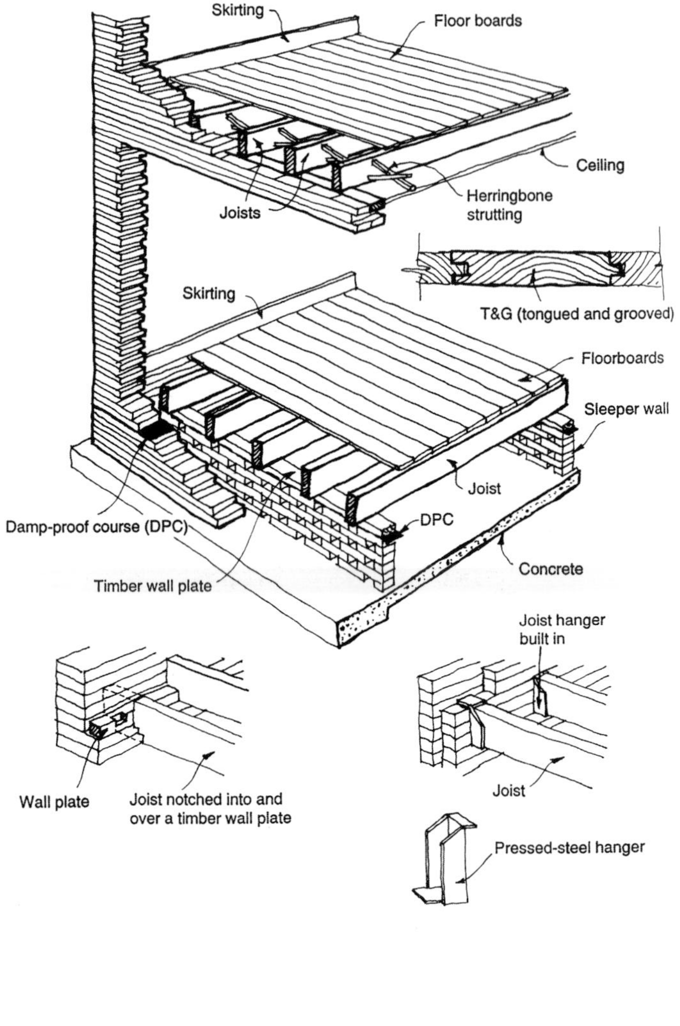

Skirting
Floor boards
Ceiling
Herringbone strutting
Joists
T&G (tongued and grooved)
Skirting
Floorboards
Sleeper wall
Joist
DPC
Damp-proof course (DPC)
Concrete
Timber wall plate
Joist hanger built in
Wall plate
Joist notched into and over a timber wall plate
Joist
Pressed-steel hanger

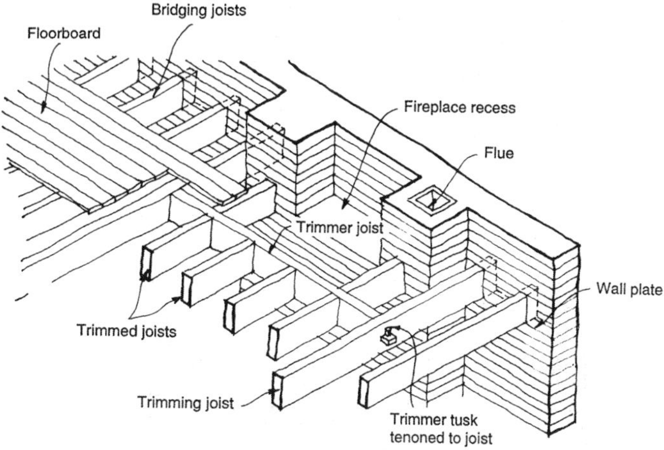
Bridging joists
Floorboard
Fireplace recess
Flue
Trimmer joist
Wall plate
Trimmed joists
Trimming joist
Trimmer tusk
tenoned to joist

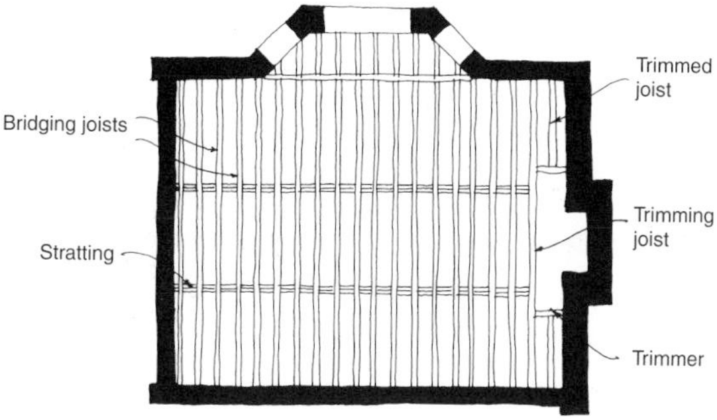
Trimmed
joist
Bridging joists
Trimming
joist
Stratting
Trimmer

Single joisted floor plan

Trimming to timber floors

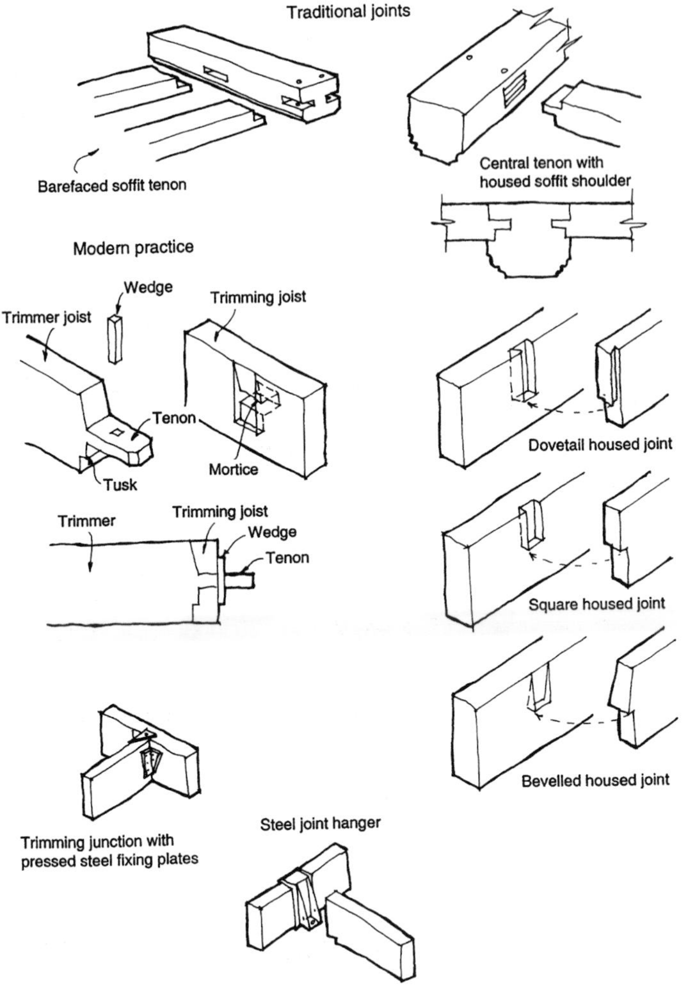
Traditional joints
Barefaced soffit tenon
Central tenon with
housed soffit shoulder
Modern practice
Wedge
Trimmer joist
Trimming joist
Tenon
Mortice
Tusk
Trimmer
Trimming joist
Wedge
Tenon
Dovetail housed joint
Square housed joint
Bevelled housed joint
Trimming junction with
pressed steel fixing plates
Steel joint hanger

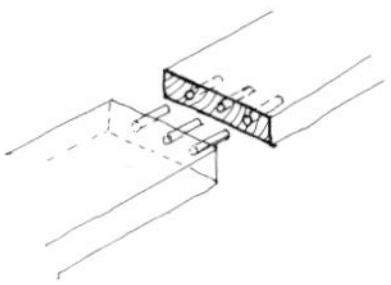

Dowelled butt joint

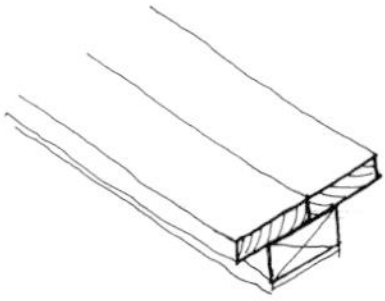

Butt joint on bearers

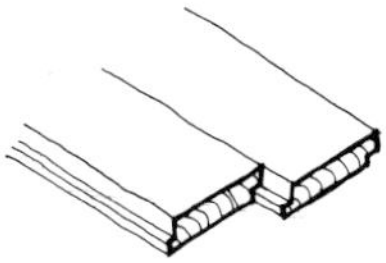

Rebated joint

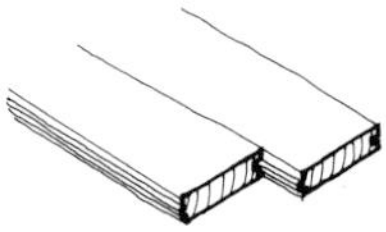

Tonove and grooved (T and G)

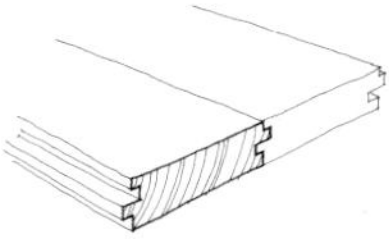

Double T and G

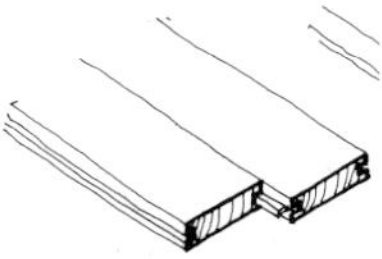

T and G with loose tongue

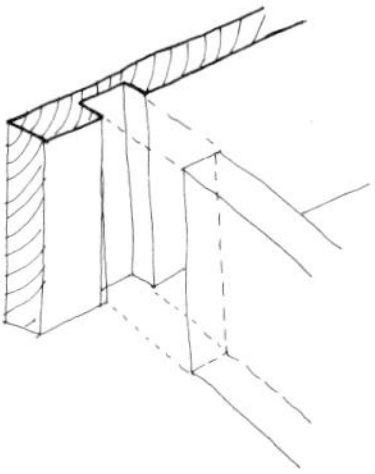

Square housing

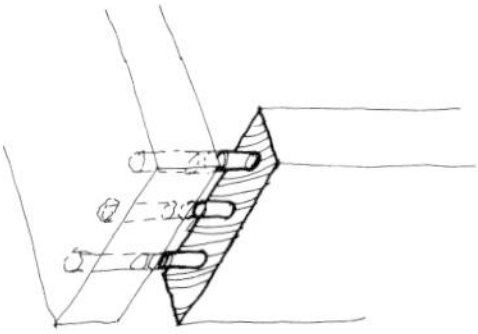

Dowelled

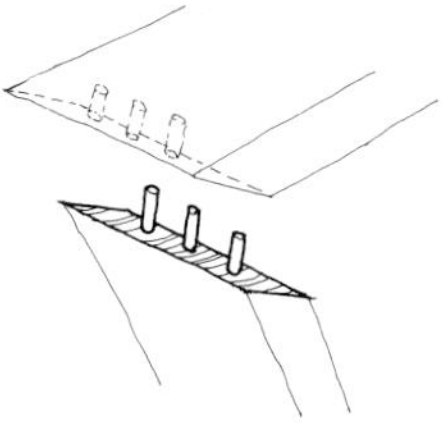

Dowlled mitred

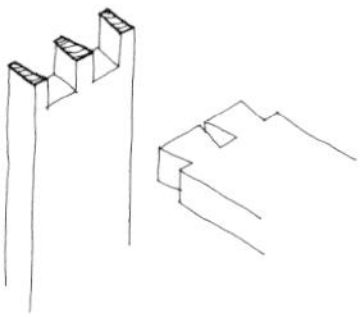

Common dovetail

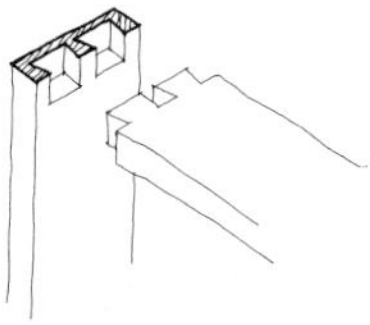

Stopped or lapped dovetail

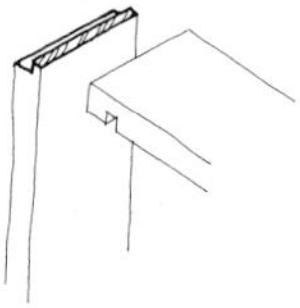

Lapped and tongue joint

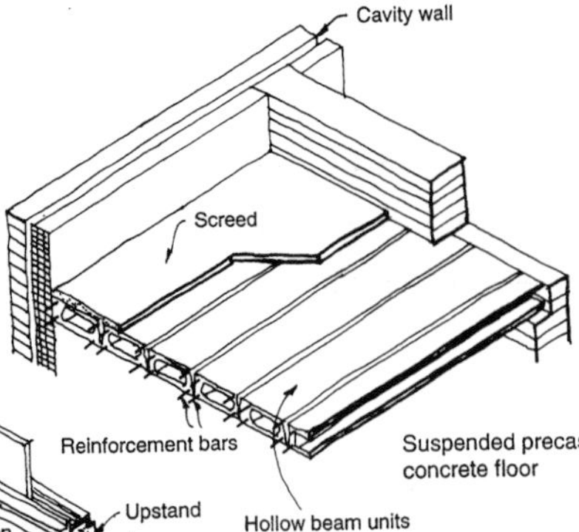

Suspended precast concrete floor

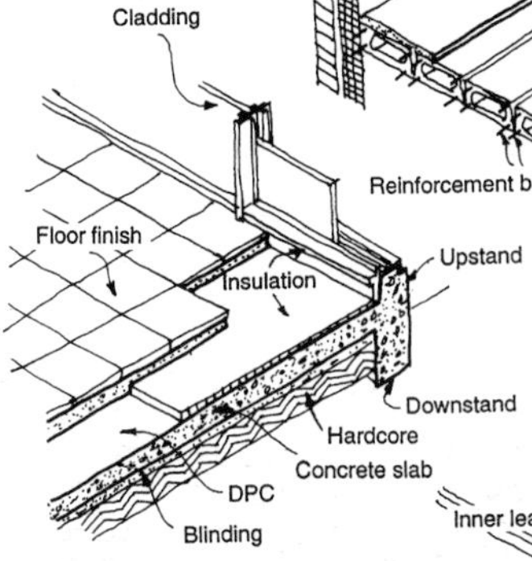

Solid ground floor

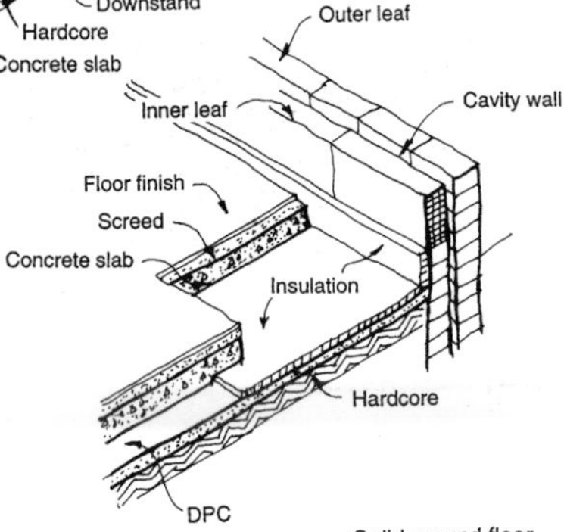

Solid ground floor

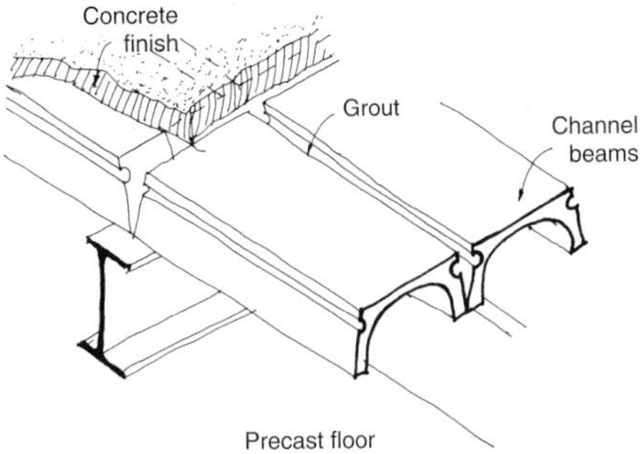

Precast floor

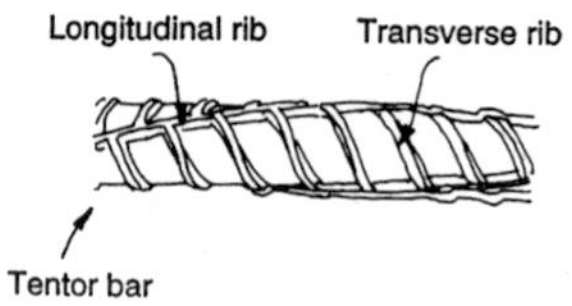

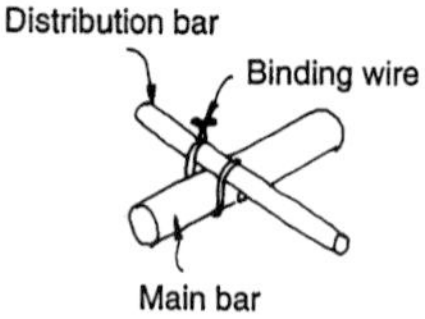

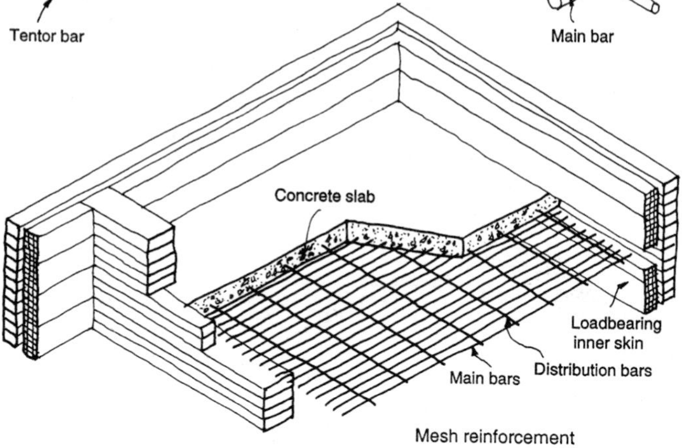

Mesh reinforcement

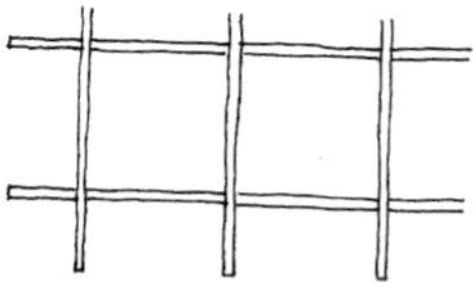

Square mesh

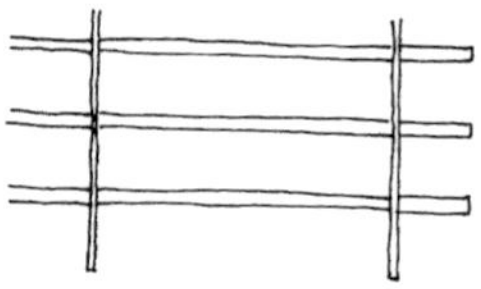

Oblong mesh

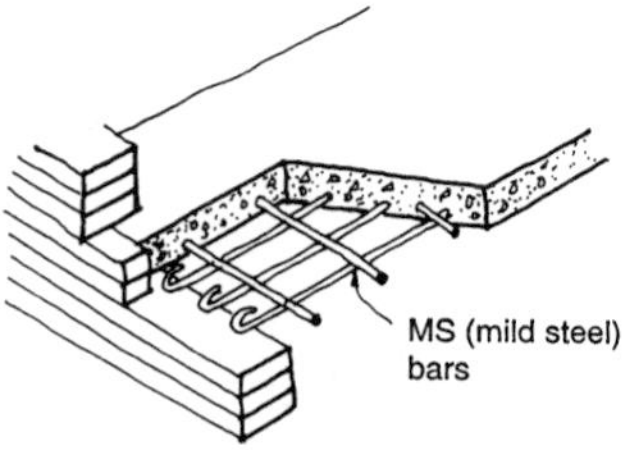

Bar reinforcement

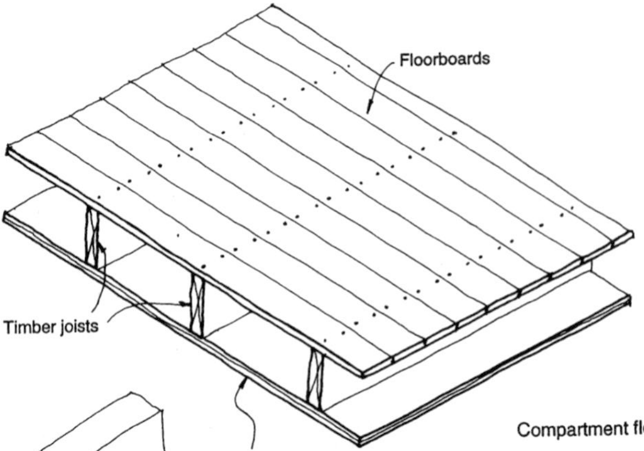
Floorboards
Timber joists
Plasterboard nailed or
screw-fixed to the timber joist

Compartment floor

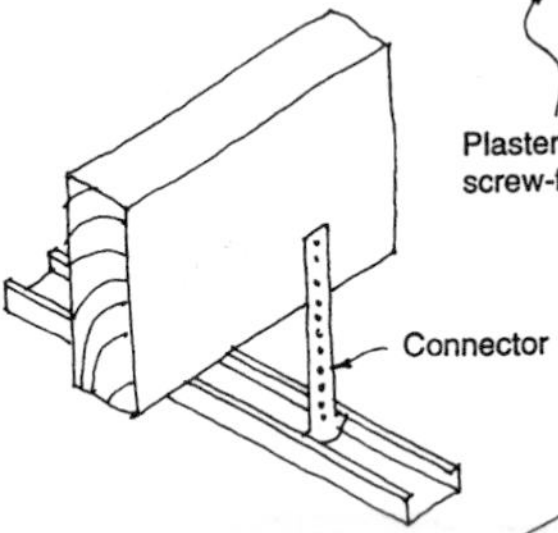
Connector

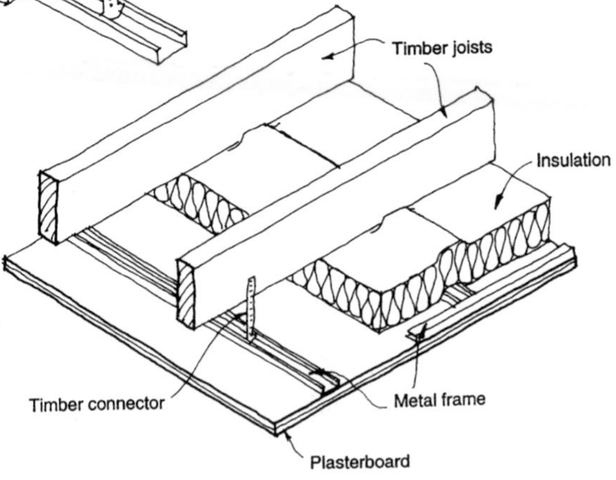
Timber joists
Insulation
Timber connector
Metal frame
Plasterboard

Liner ceiling

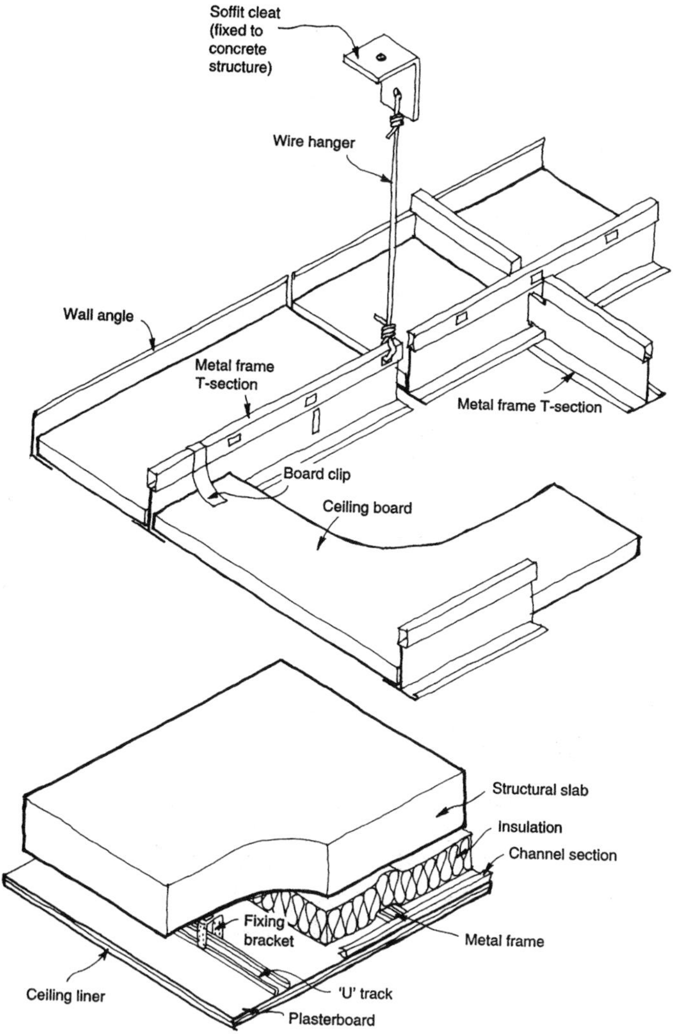
Soffit cleat
(fixed to
concrete
structure)
Wire hanger
Wall angle
Metal frame
T-section
Metal frame T-section
Board clip
Ceiling board
Structural slab
Insulation
Channel section
Fixing
bracket
Metal frame
Ceiling liner
'U' track
Plasterboard

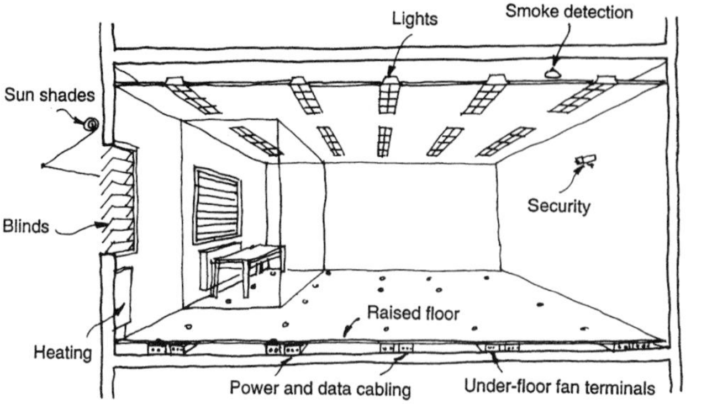

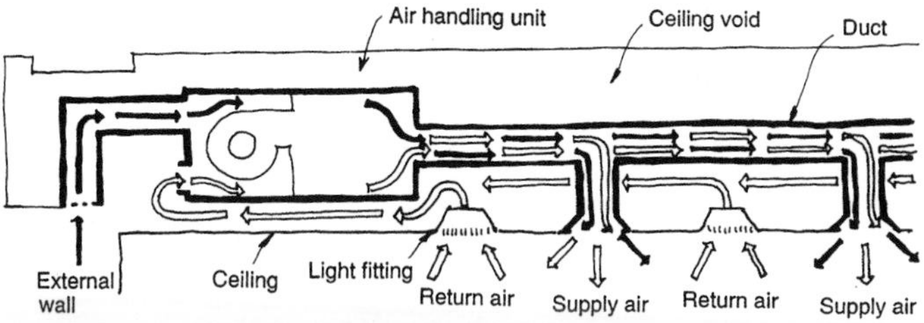

Integrated lighting and air conditioning

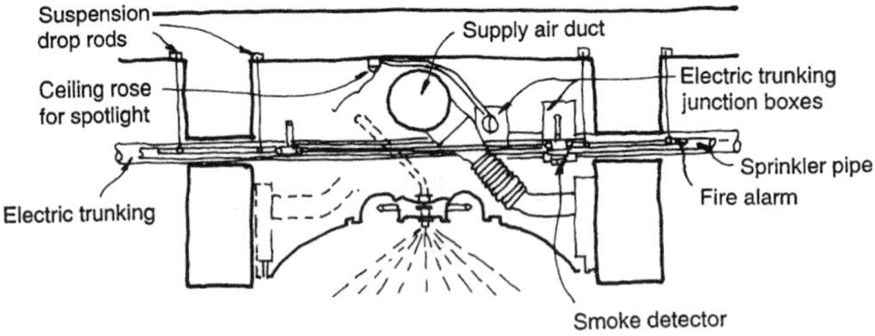

Sprinkler in integrated ceiling

Shallow void
battened floor

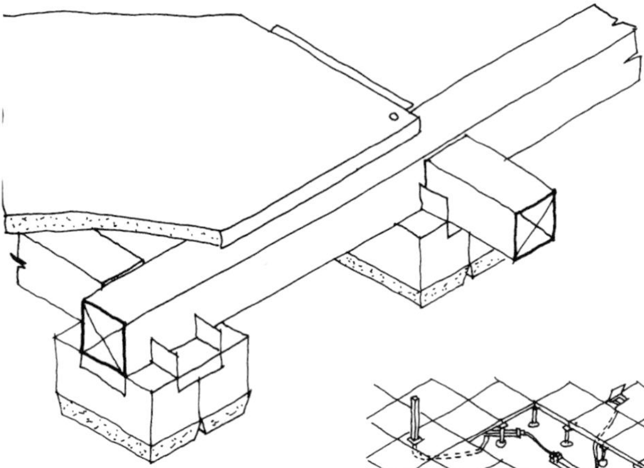

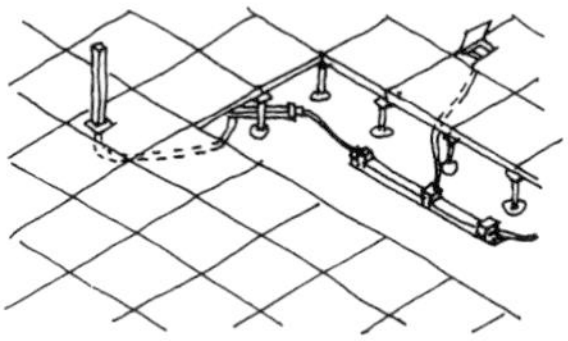

Underfloor busbar system

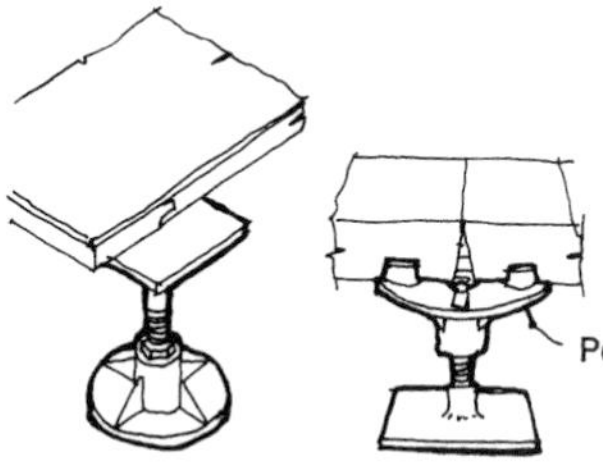

Deep void platform
floor supports

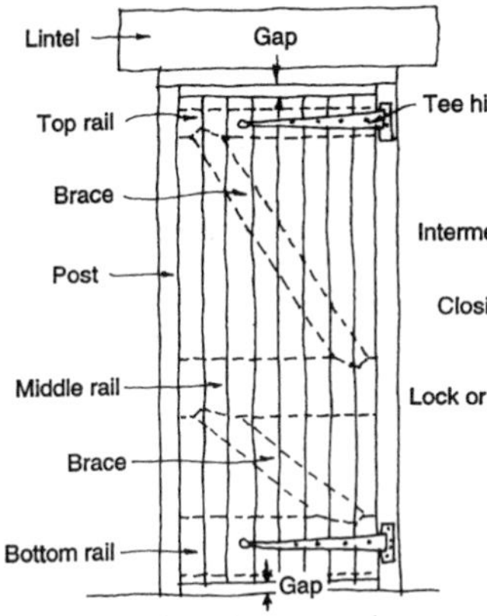

Ledged, braced and battened door

Panelled door

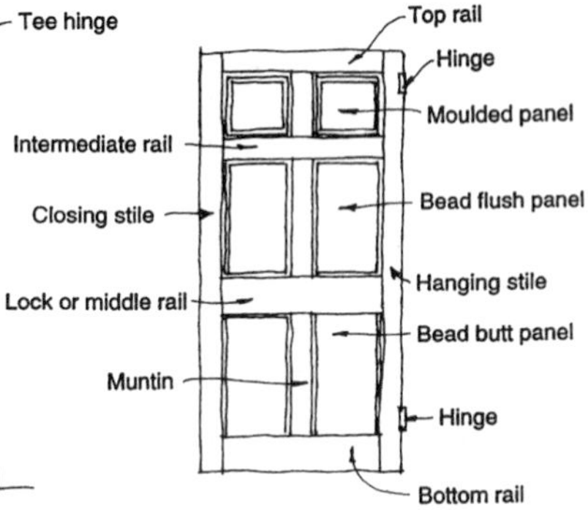

Interior flush door

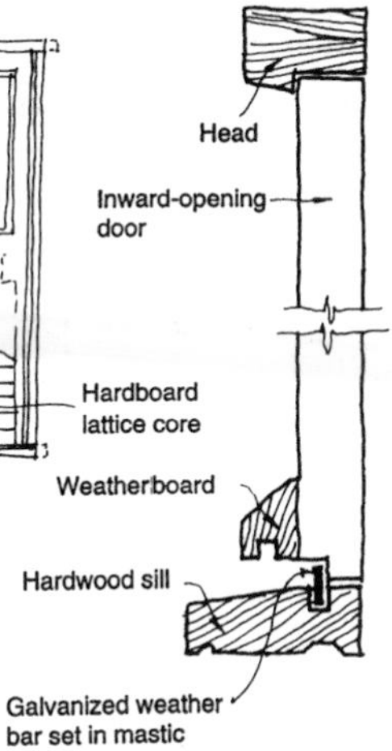

Timber door frame

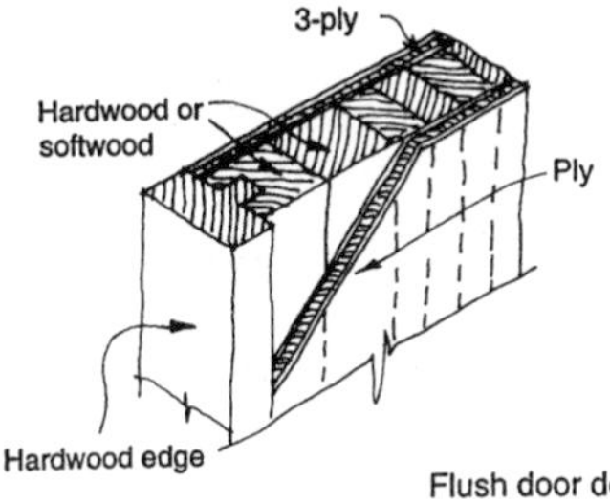

Flush door detail

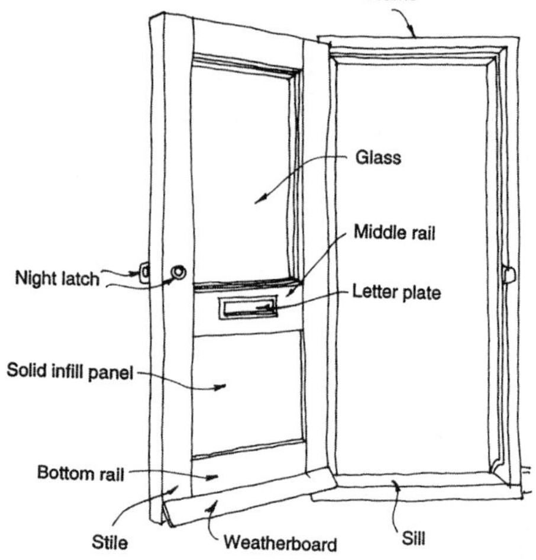

Types of hinge

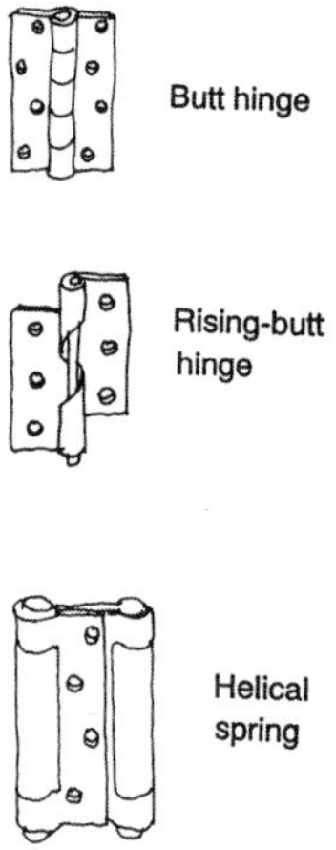

Frame fixing cramp

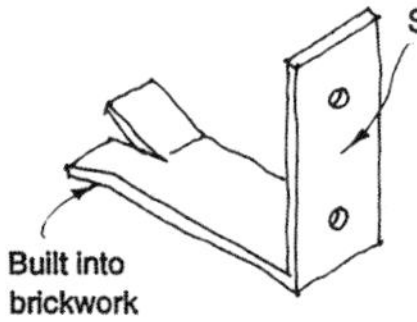

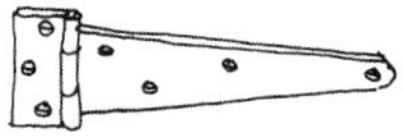

T-hinge

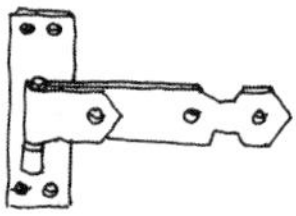

Hook and band hinge

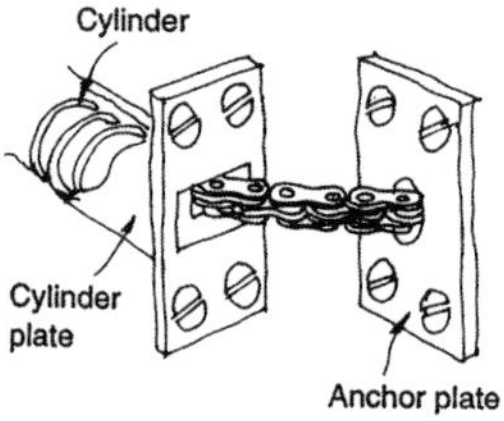

'Perko' door closer

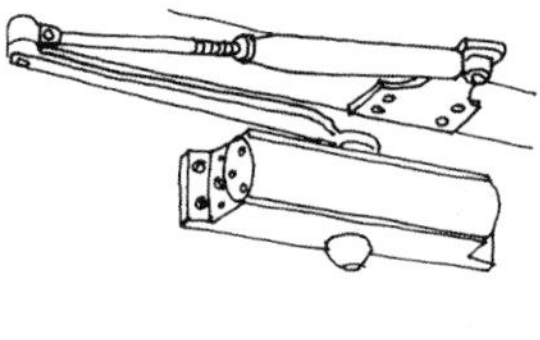

Parallel arm door closer

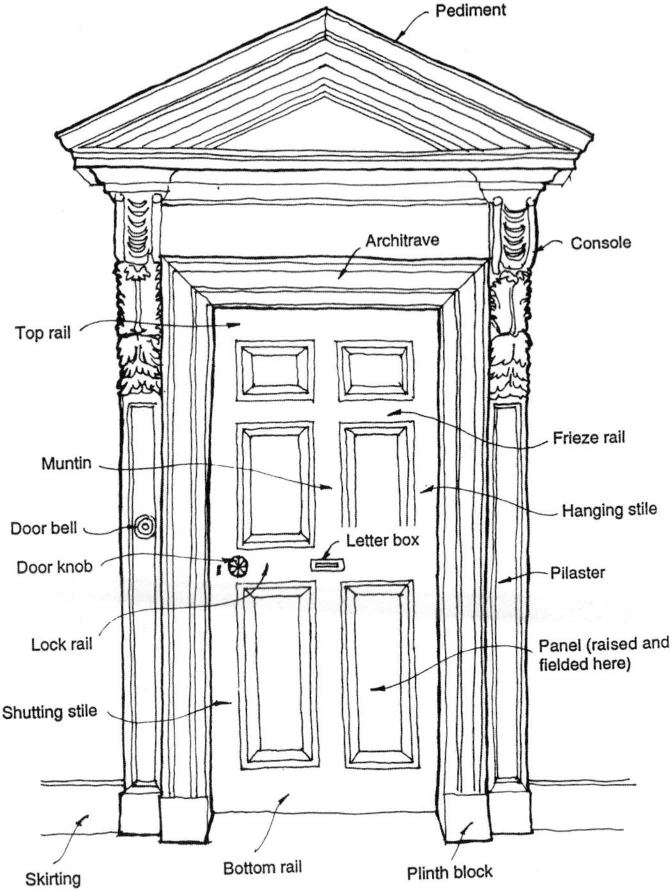
Pediment
Architrave
Console
Top rail
Frieze rail
Muntin
Hanging stile
Door bell
Letter box
Door knob
Pilaster
Lock rail
Panel (raised and fielded here)
Shutting stile
Skirting
Bottom rail
Plinth block

Traditional panel door

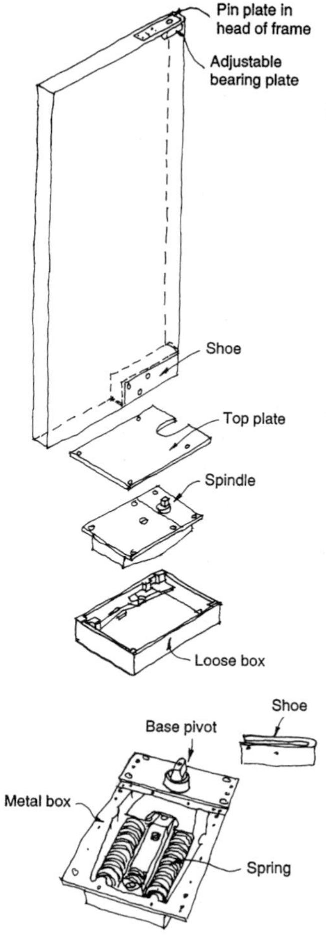

Typical installation of door spring

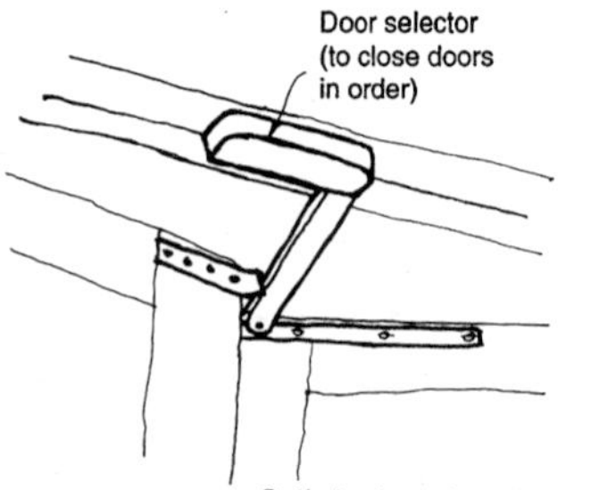

Self-closing rebated doors

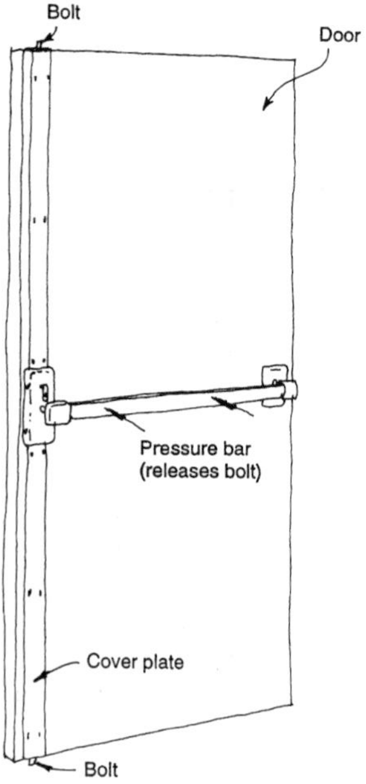

Mortice panic bolt

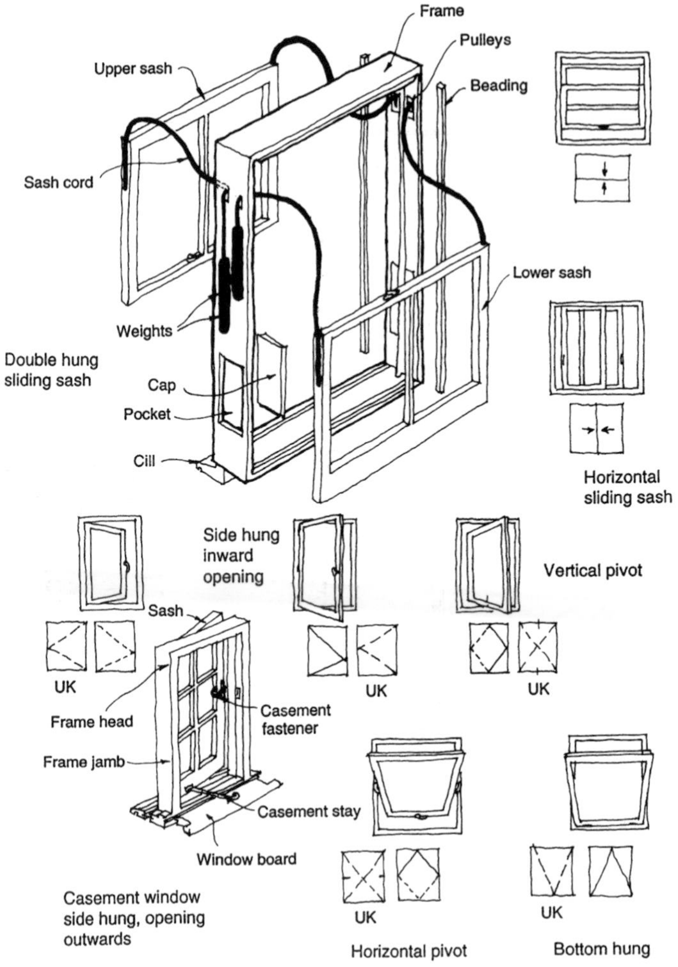
Frame
Pulleys
Upper sash
Beading
Sash cord
Lower sash
Weights
Double hung
sliding sash
Cap
Pocket
Cill
Horizontal
sliding sash
Side hung
inward
opening
Vertical pivot
Sash
UK
UK
UK
Frame head
Casement
fastener
Frame jamb
Casement stay
Window board
Casement window
side hung, opening
outwards
UK
UK
Horizontal pivot
Bottom hung

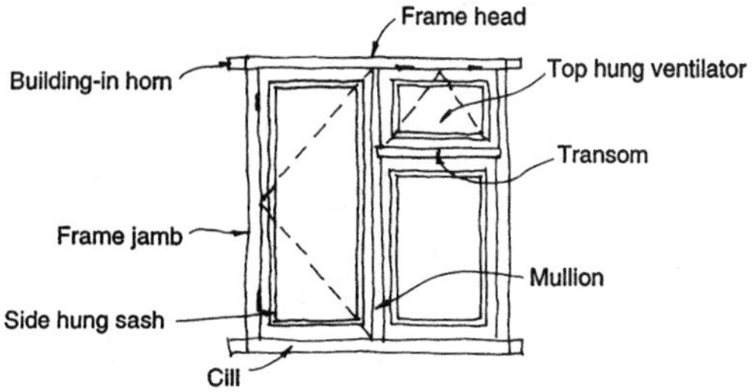
Frame head
Building-in horn
Top hung ventilator
Transom
Frame jamb
Mullion
Side hung sash
Cill

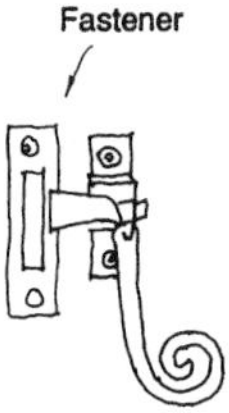
Fastener

Timber casement window

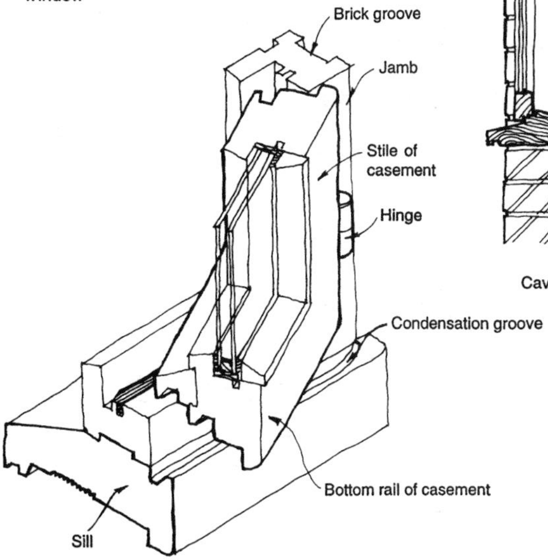
Brick groove
Jamb
Stile of casement
Hinge
Condensation groove
Bottom rail of casement
Sill

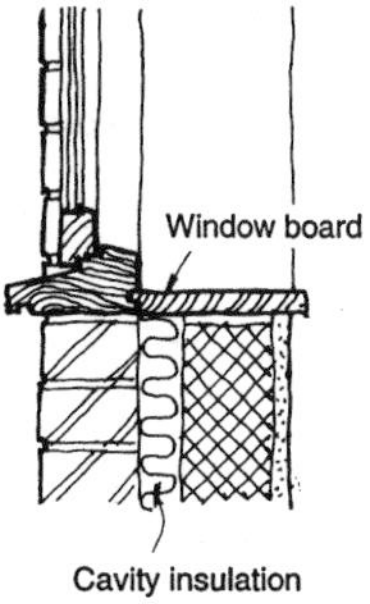
Window board
Cavity insulation

Horizontal mortice lock

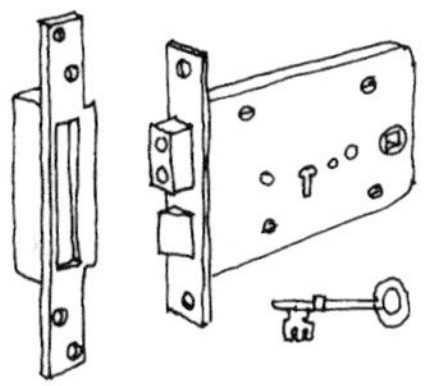

Cylinder rim latch

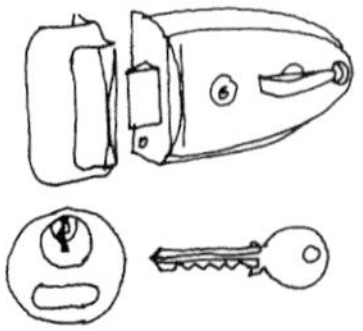

Knob set

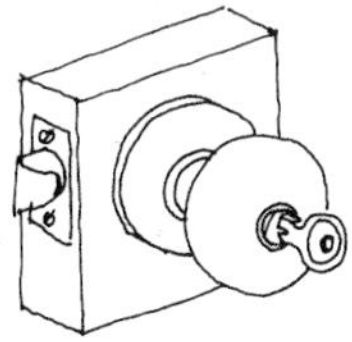

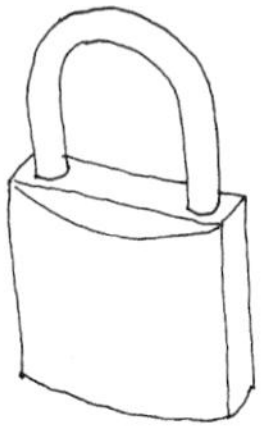

Padlock

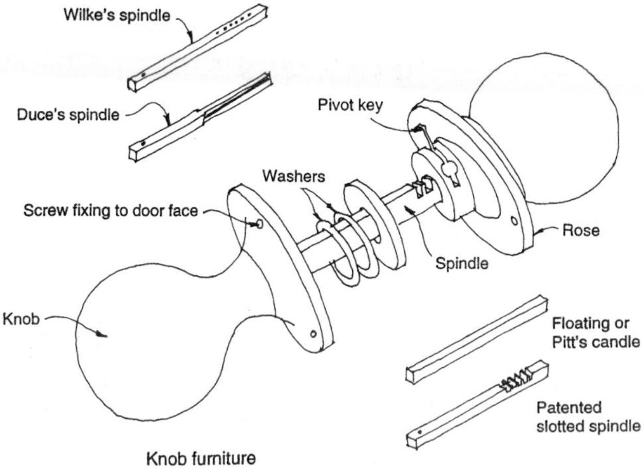

Knob furniture

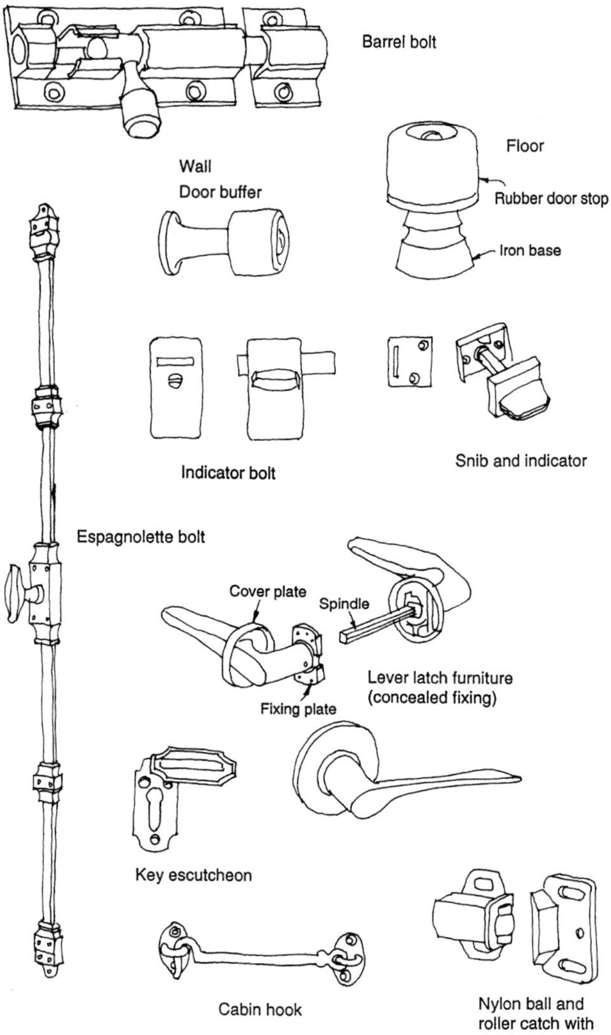
Barrel bolt
Floor
Wall
Door buffer
Rubber door stop
Iron base
Indicator bolt
Snib and indicator
Espagnolette bolt
Cover plate
Spindle
Lever latch furniture
(concealed fixing)
Fixing plate
Key escutcheon
Cabin hook
Nylon ball and
roller catch with
steel spring

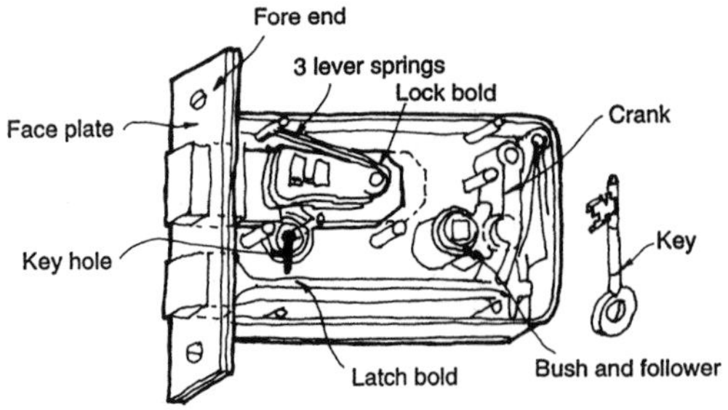

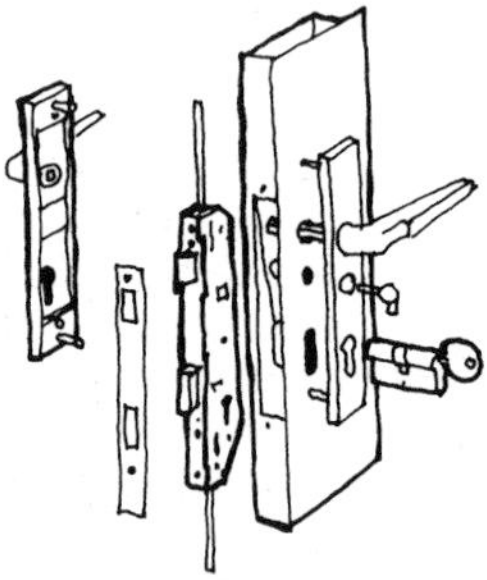

Multi-point lock

Cylinder locks

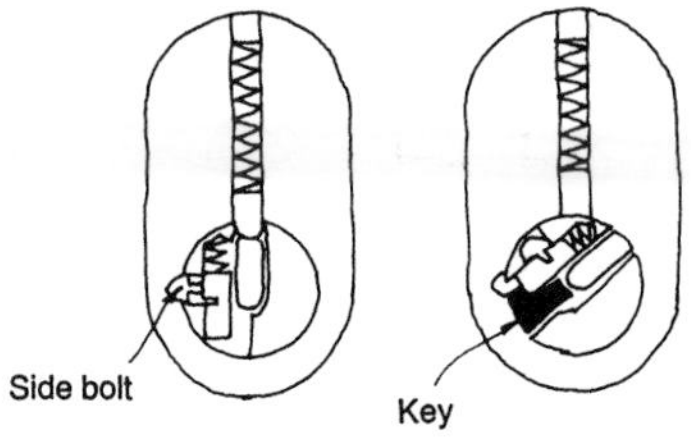

Springloaded blocking pins lock

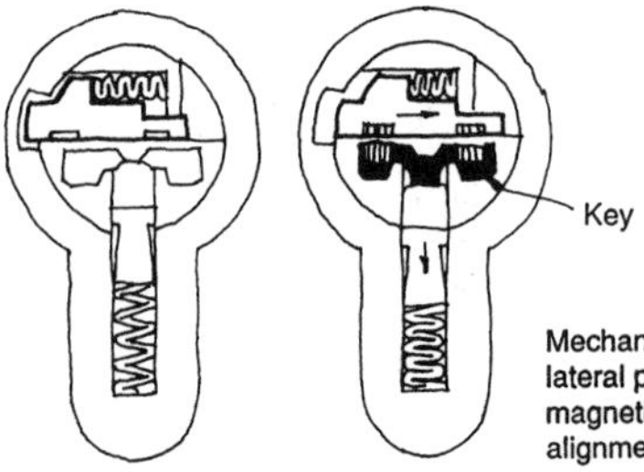

Mechanical and lateral pins magnetic alignment

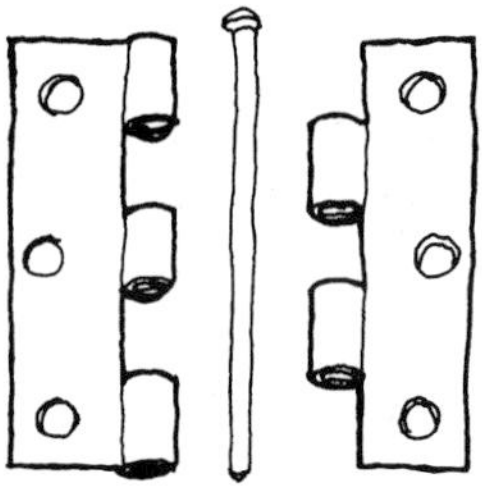

Loose pin

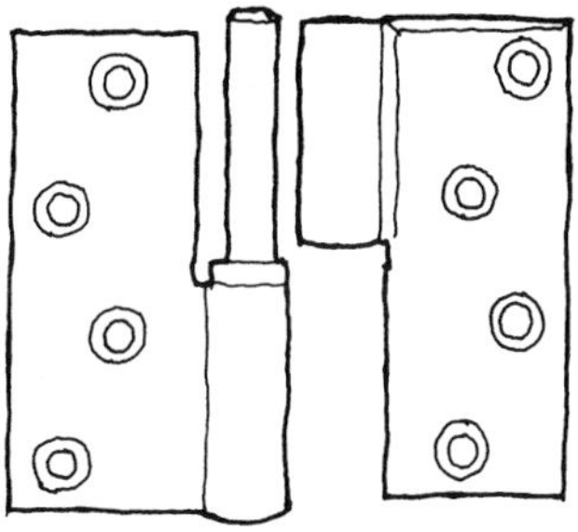

Lift-off butt

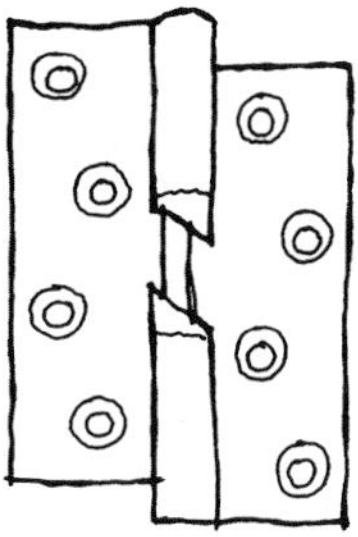

Rising butt

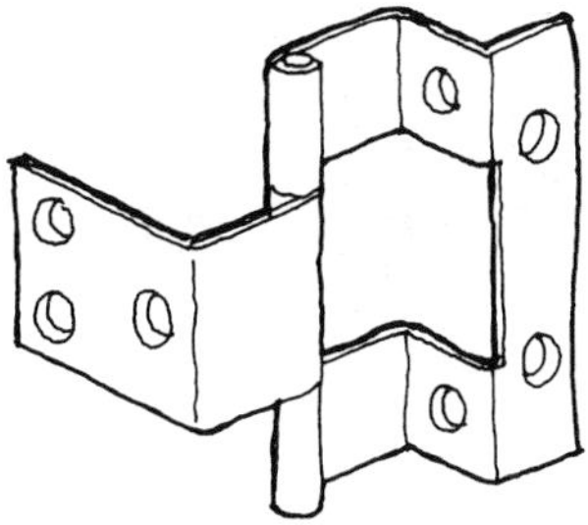

Cranked

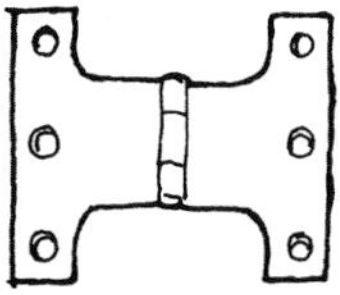

Parliament

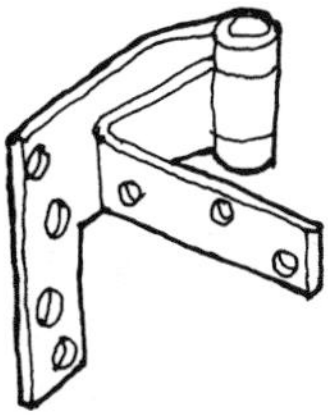

Offset-easyclean

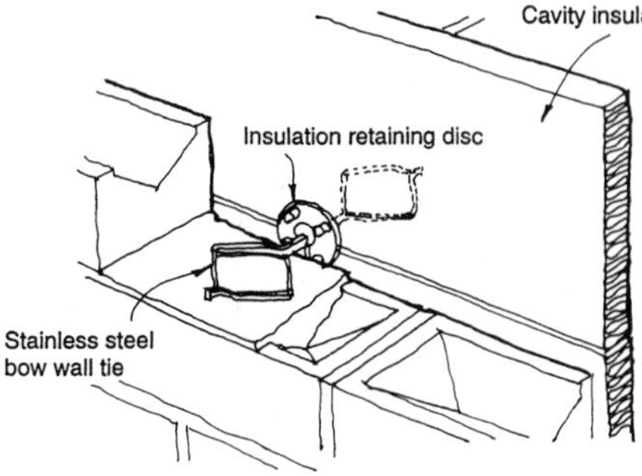
Cavity insulation board
Insulation retaining disc
Stainless steel
bow wall tie

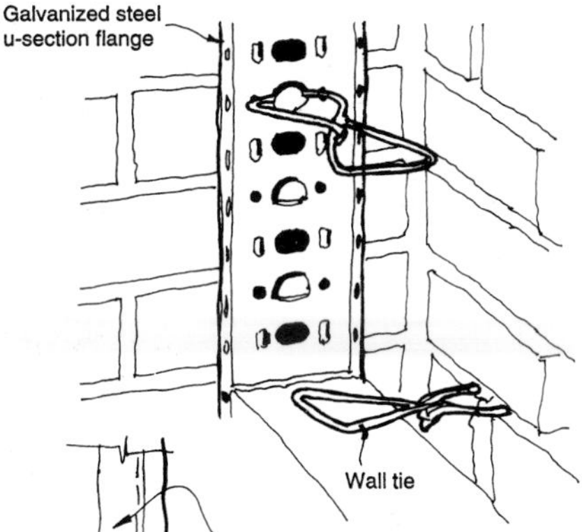
Galvanized steel
u-section flange
Wall tie

Cladding fixing

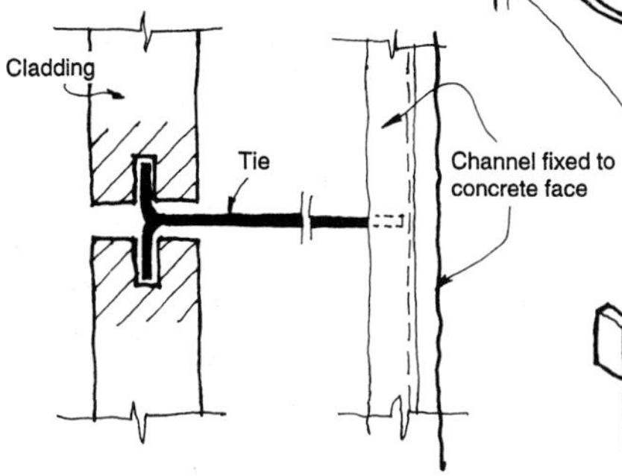
Cladding
Tie
Channel fixed to
concrete face

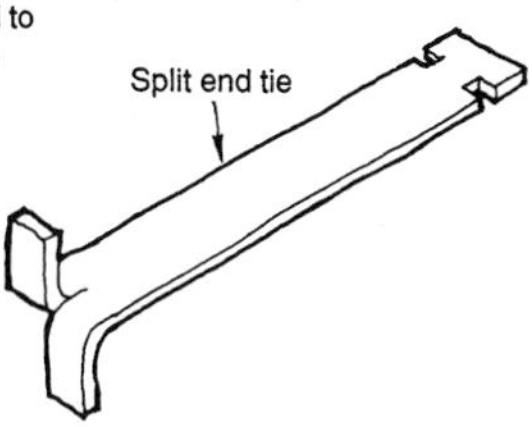
Split end tie

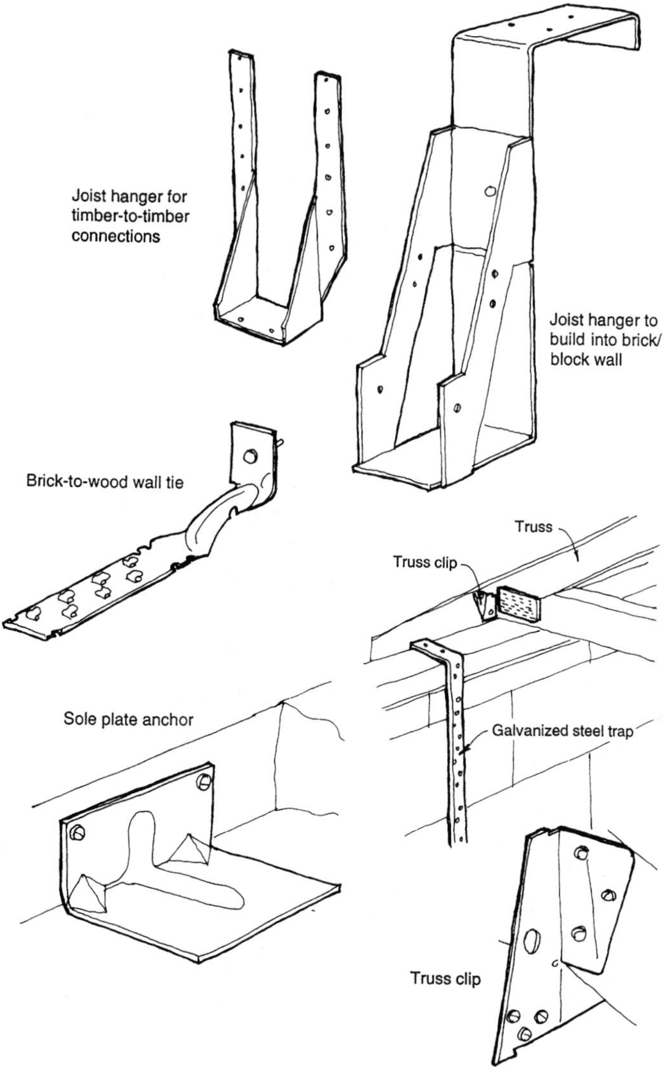
Joist hanger for
timber-to-timber
connections
Joist hanger to
build into brick/
block wall
Brick-to-wood wall tie
Truss
Truss clip
Galvanized steel trap
Sole plate anchor
Truss clip

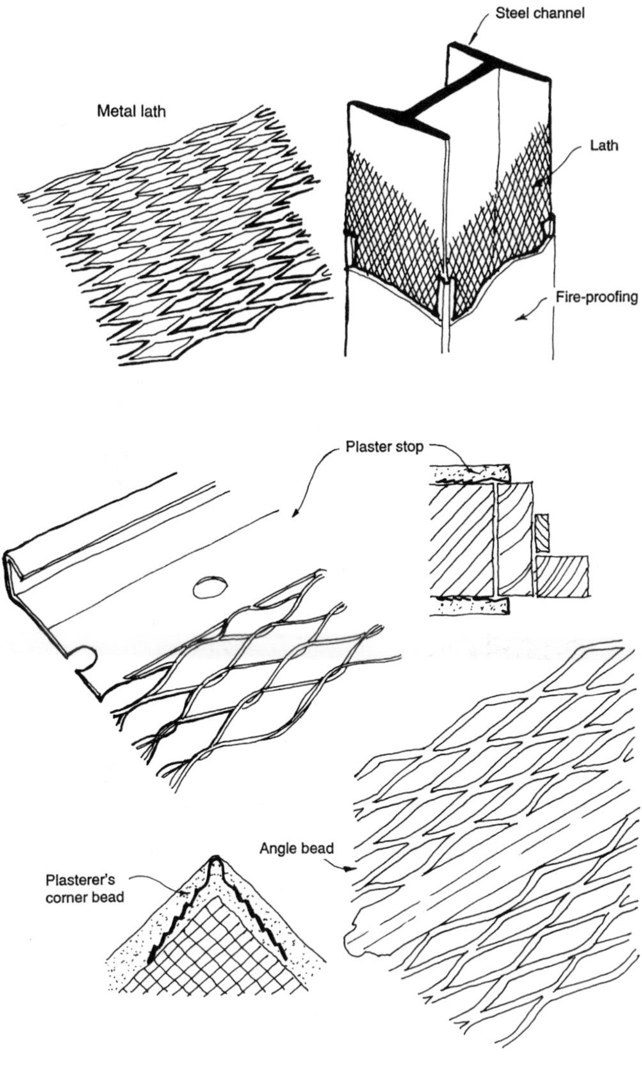
Steel channel
Metal lath
Lath
Fire-proofing
Plaster stop
Angle bead
Plasterer's
corner bead

Steel lintel
for cavity wall

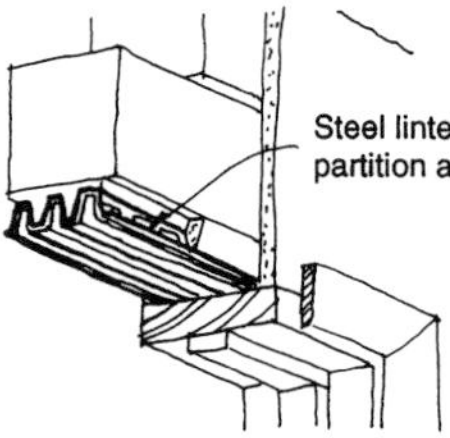

Steel lintel supporting
partition above door

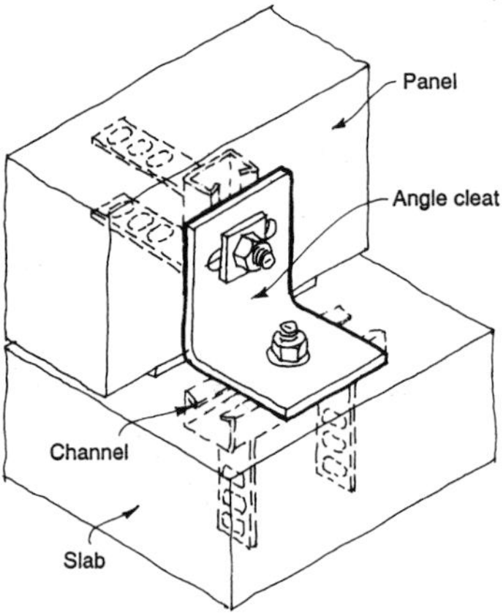

Fixing for precast
cladding panels

Nails

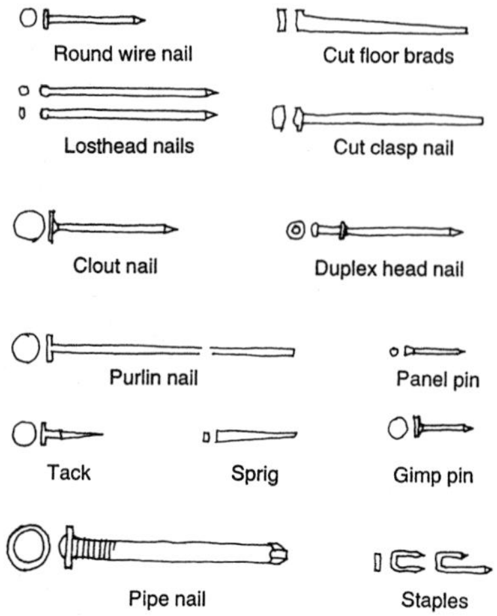

Screws

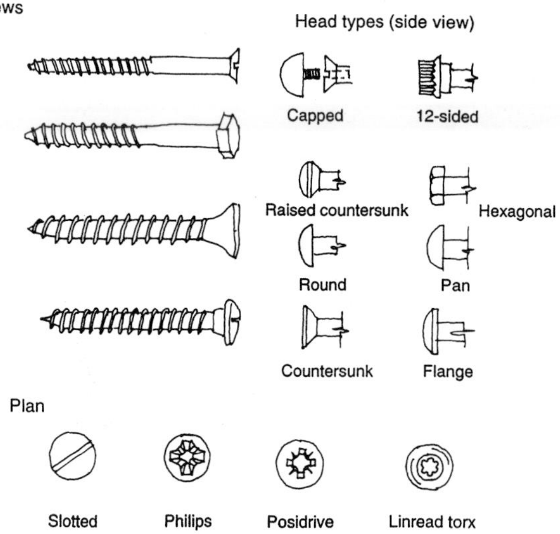

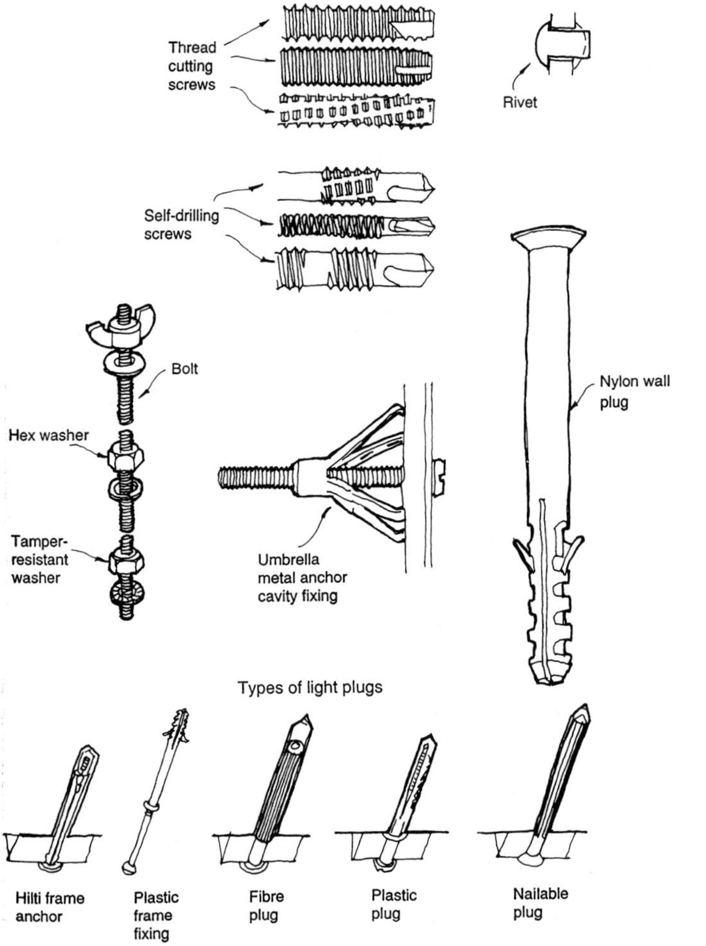
Thread
cutting
screws
Rivet
Self-drilling
screws
Bolt
Nylon wall
plug
Hex washer
Tamper-
resistant
washer
Umbrella
metal anchor
cavity fixing
Types of light plugs
Hilti frame
anchor
Plastic
frame
fixing
Fibre
plug
Plastic
plug
Nailable
plug

Plasterwork tools

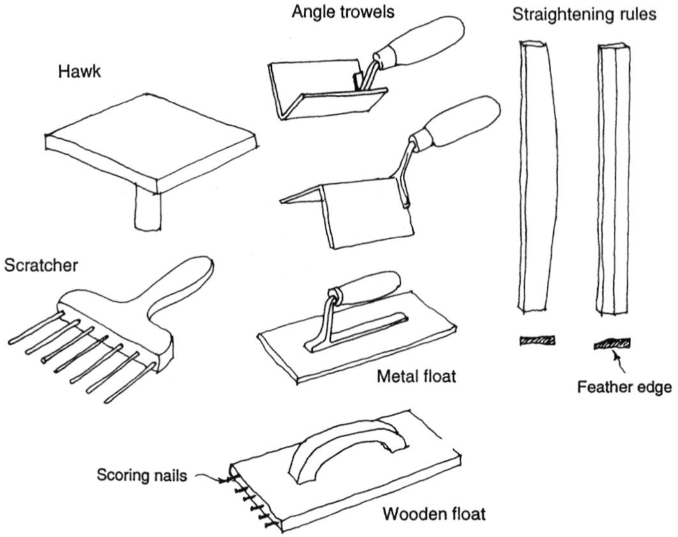
Angle trowels
Straightening rules
Hawk
Scratcher
Metal float
Feather edge
Scoring nails
Wooden float

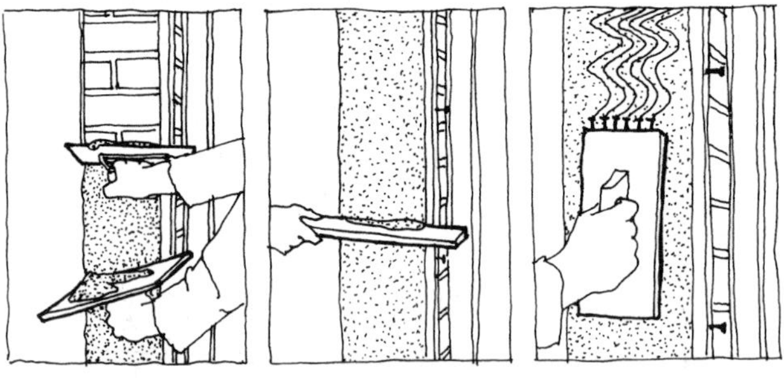
Applying first
coat of plaster
Levelling
(ruling out)
Scoring

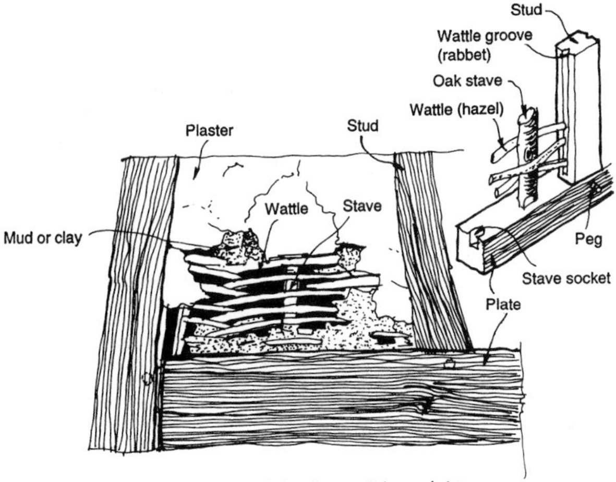

*Wattle is daubed both sides with a mixture of clay, dung and chopped straw

Lath and plaster ceiling

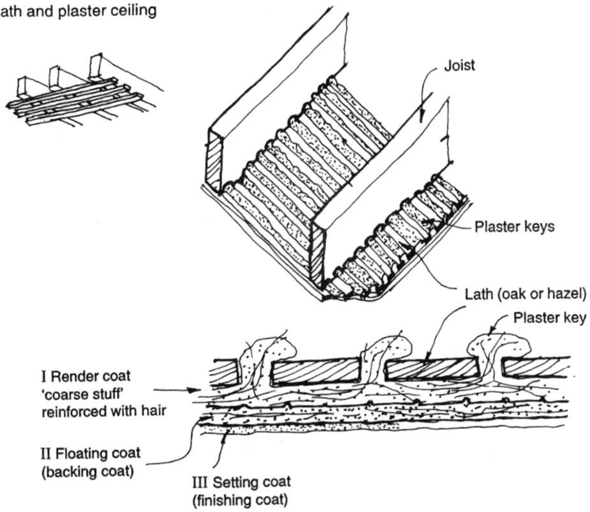

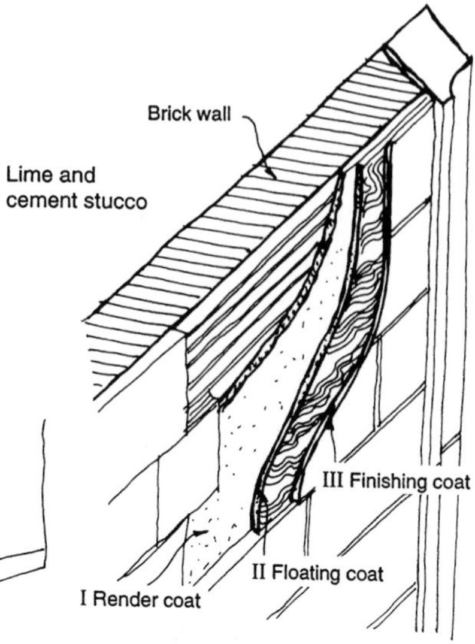
Brick wall
Lime and
cement stucco
III Finishing coat
II Floating coat
I Render coat

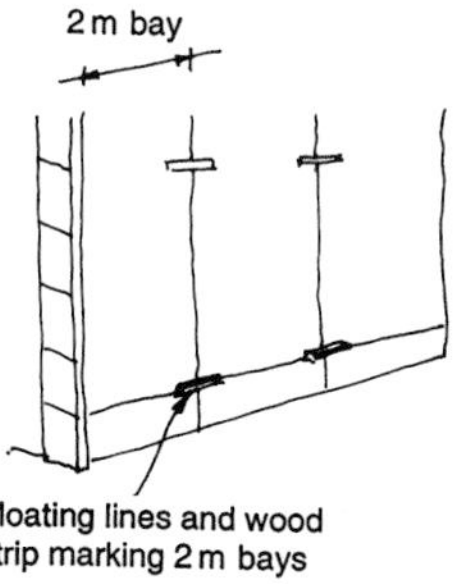
2 m bay
Floating lines and wood
strip marking 2 m bays

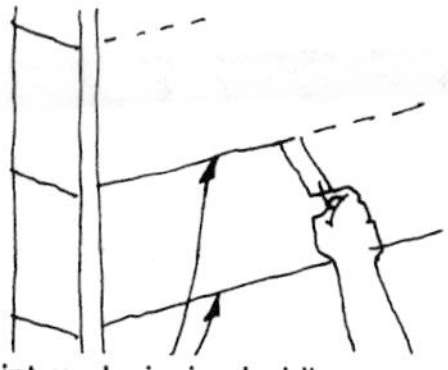
Joint marks incised while
finishing coat still soft

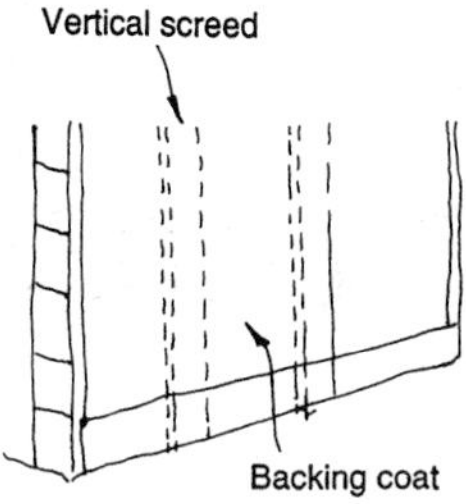
Vertical screed
Backing coat

Running a cornice in situ

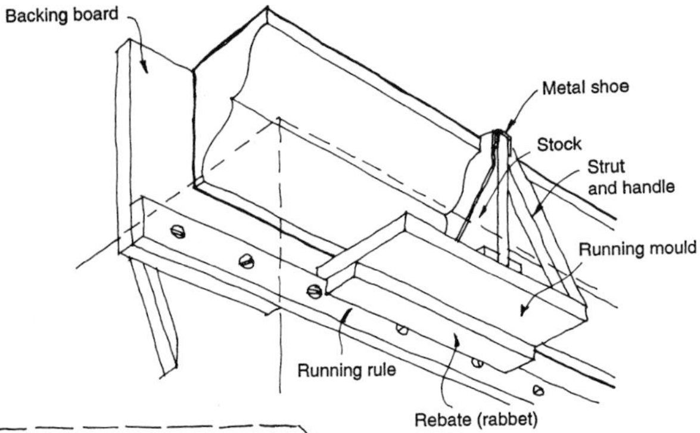

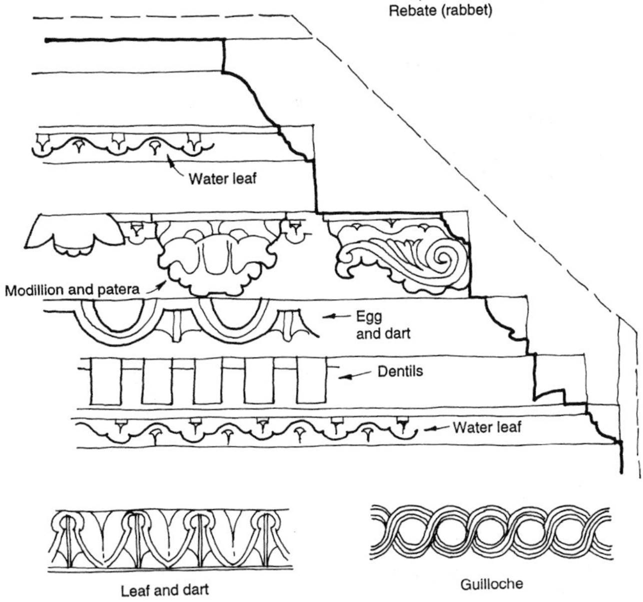

Cornice ornamentation (planted in wet stucco)

Plasterboard fixed directly to wall

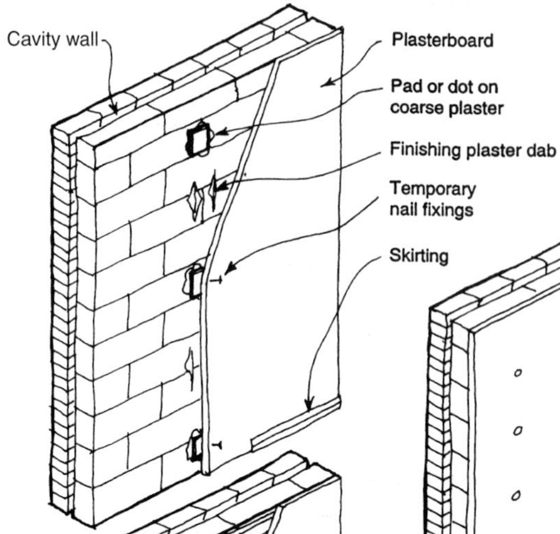

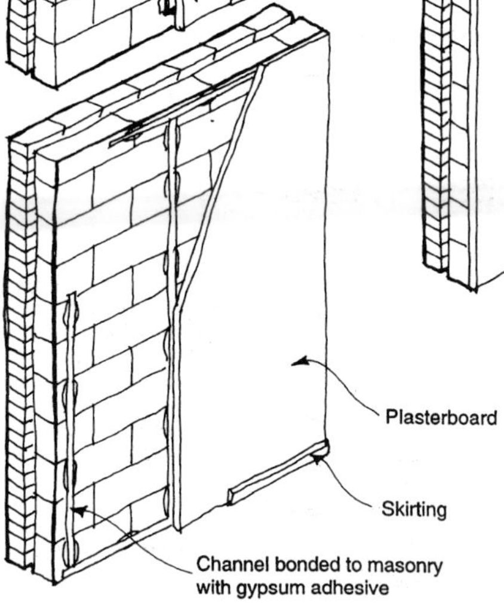

Plasterboard fixed to lightweight metal furring channels

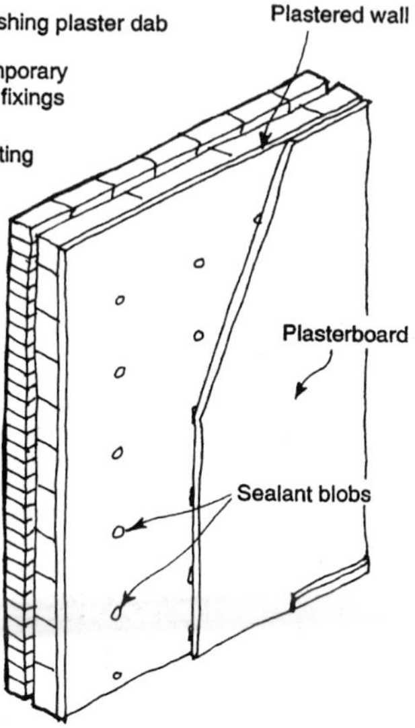

Plasterboard directly on plastered wall

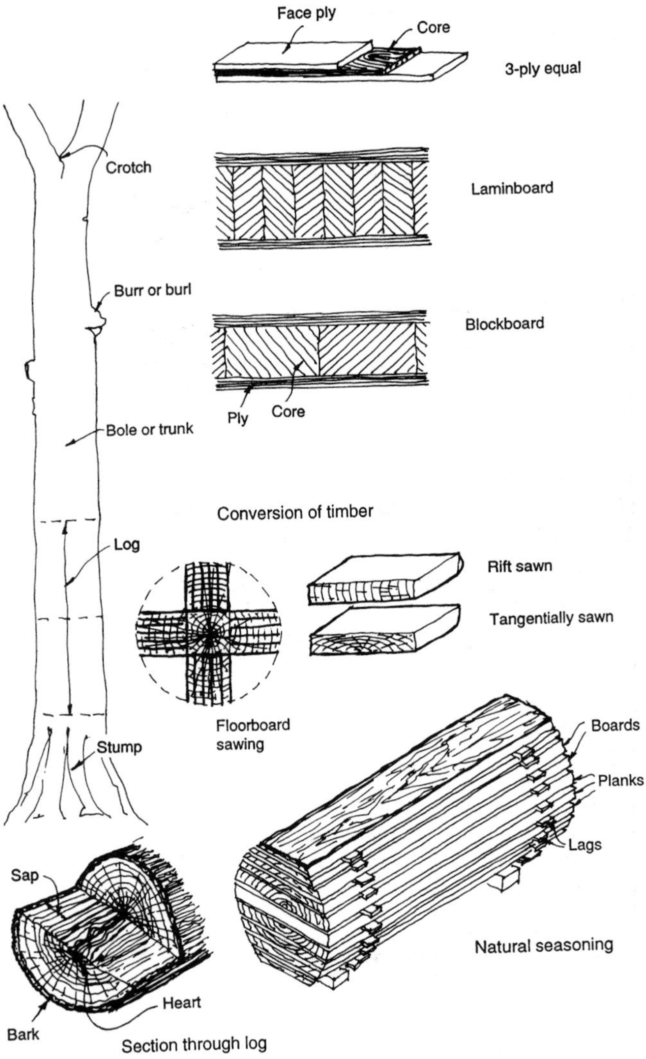
Face ply
Core
3-ply equal
Crotch
Laminboard
Burr or burl
Blockboard
Bole or trunk
Ply
Core
Conversion of timber
Log
Rift sawn
Tangentially sawn
Floorboard sawing
Stump
Boards
Planks
Lags
Sap
Natural seasoning
Heart
Bark
Section through log

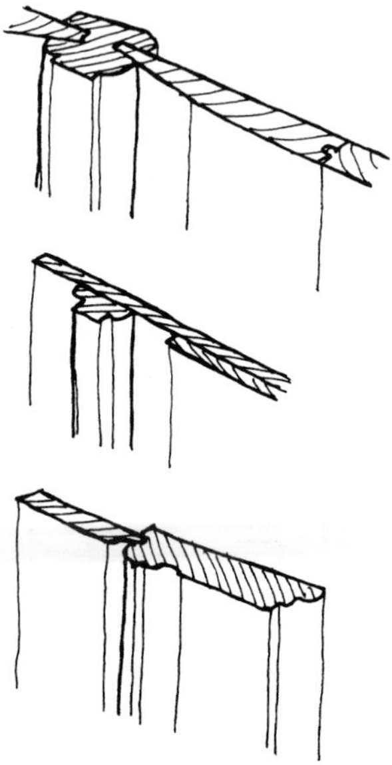

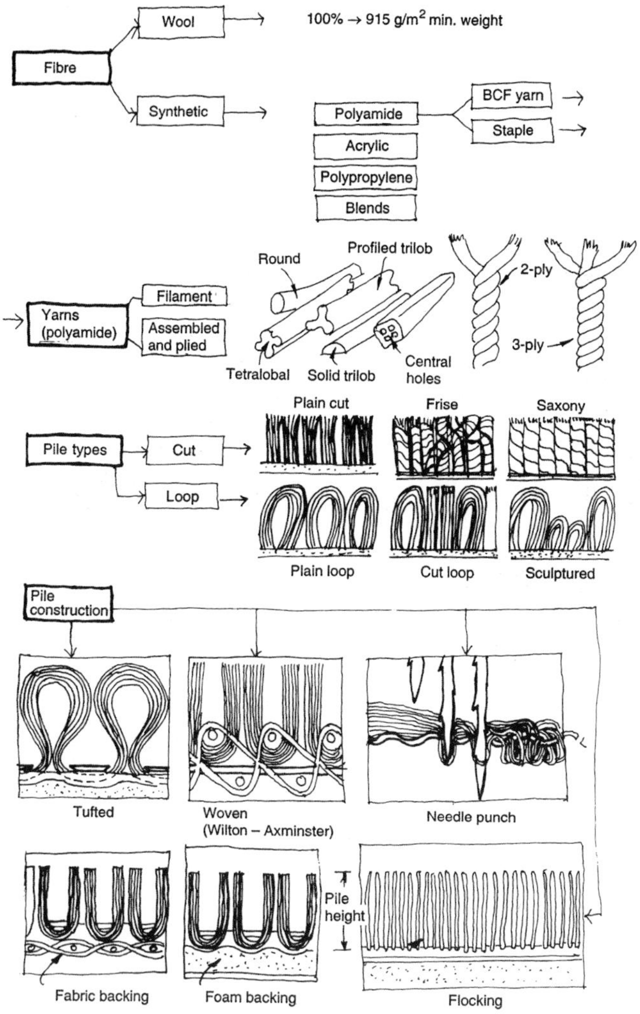
Wool
100% → 915 g/m² min. weight
Fibre
Synthetic
Polyamide
Acrylic
Polypropylene
Blends
BCF yarn
Staple
Yarns (polyamide)
Filament
Assembled and plied
Round
Profiled trilob
Tetralobal
Solid trilob
Central holes
2-ply
3-ply
Pile types
Cut
Loop
Plain cut
Frise
Saxony
Plain loop
Cut loop
Sculptured
Pile construction
Tufted
Woven
(Wilton – Axminster)
Needle punch
Pile height
Fabric backing
Foam backing
Flocking

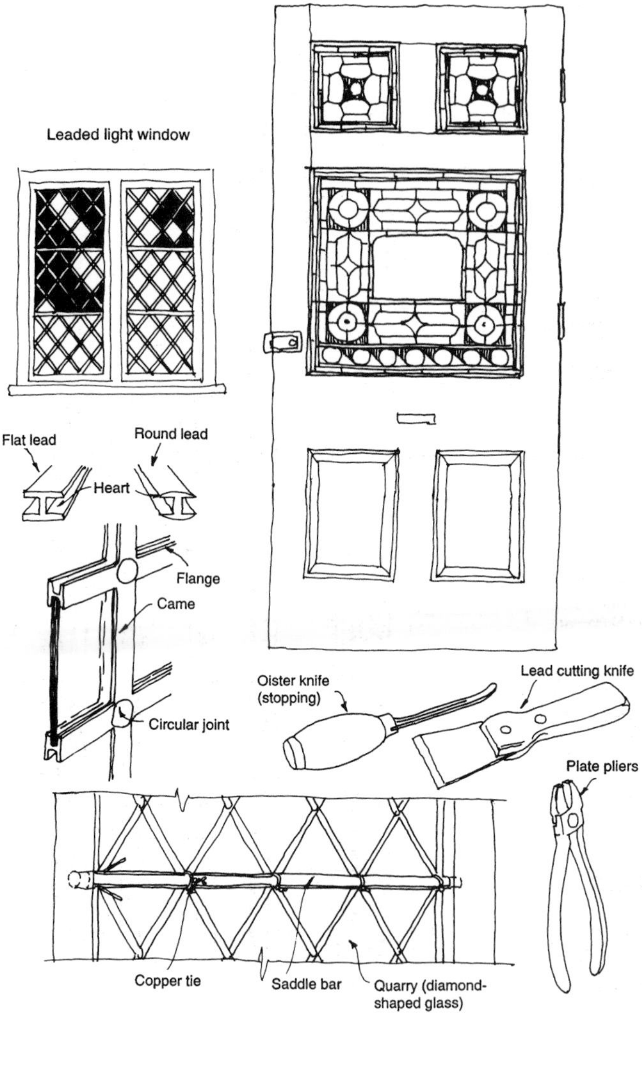
Leaded light window
Flat lead
Round lead
Heart
Flange
Came
Circular joint
Oister knife
(stopping)
Lead cutting knife
Plate pliers
Copper tie
Saddle bar
Quarry (diamond-
shaped glass)

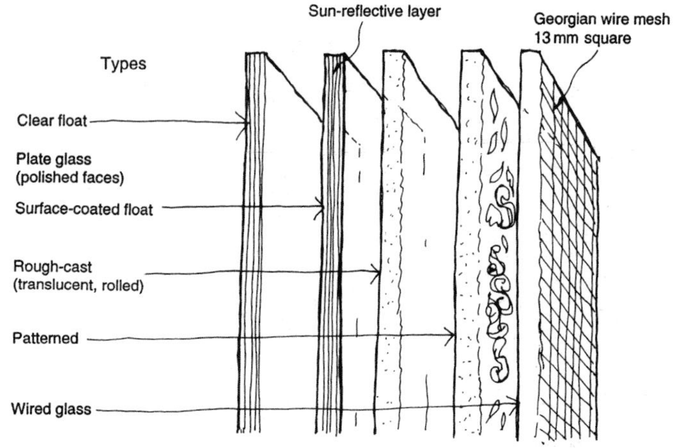
Sun-reflective layer
Georgian wire mesh 13 mm square
Types
Clear float
Plate glass (polished faces)
Surface-coated float
Rough-cast (translucent, rolled)
Patterned
Wired glass

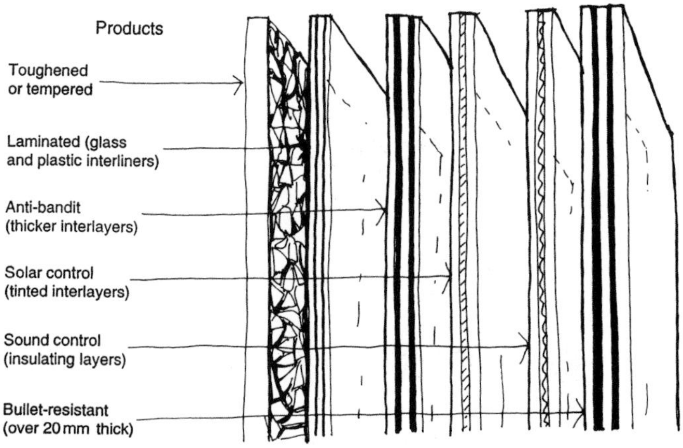
Products
Toughened or tempered
Laminated (glass and plastic interliners)
Anti-bandit (thicker interlayers)
Solar control (tinted interlayers)
Sound control (insulating layers)
Bullet-resistant (over 20 mm thick)

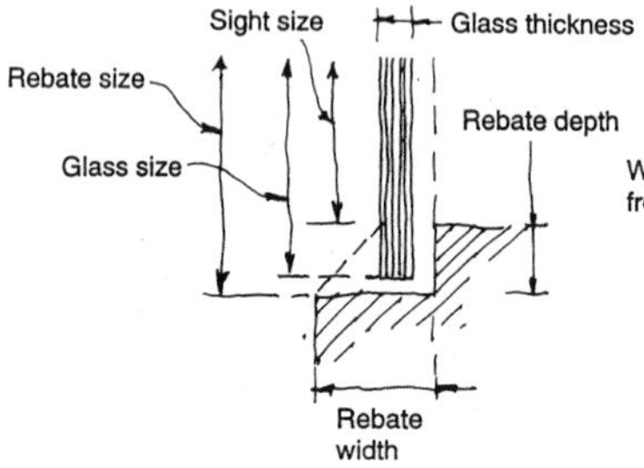

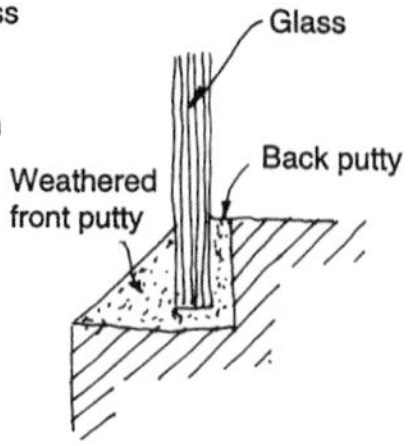

Putty glazing

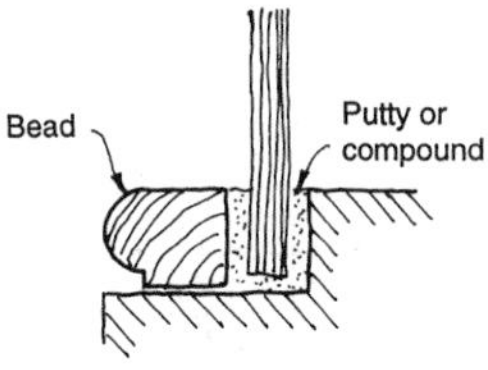

Bead glazing

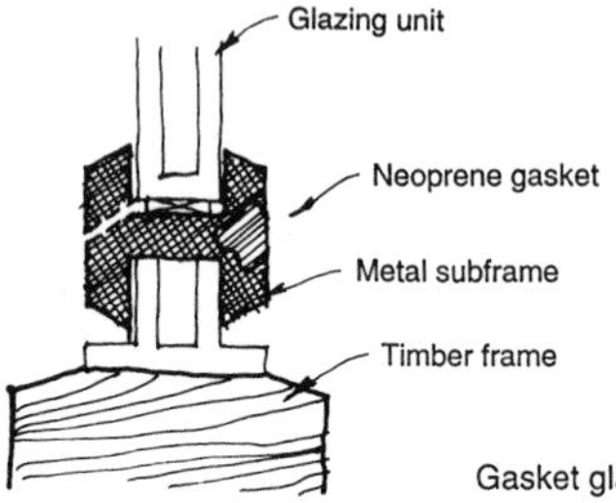

Gasket glazing

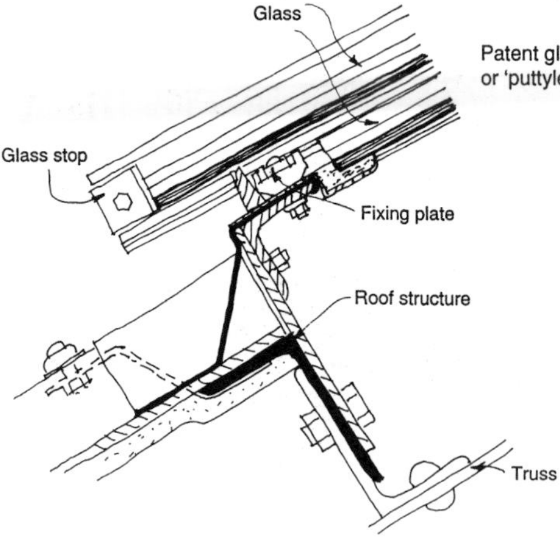

Patent glazing
or 'puttyless' glazing

Cesspool

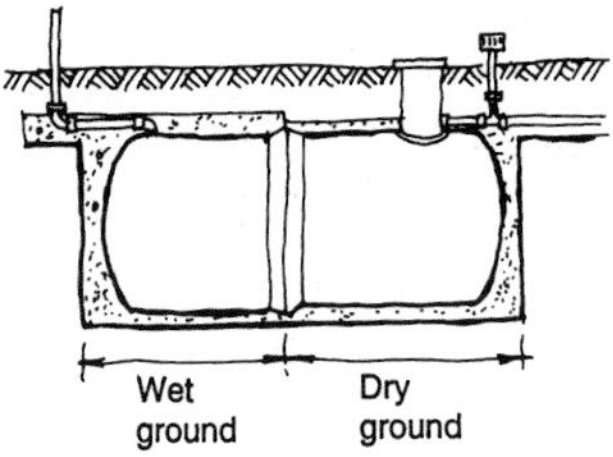

Septic tank

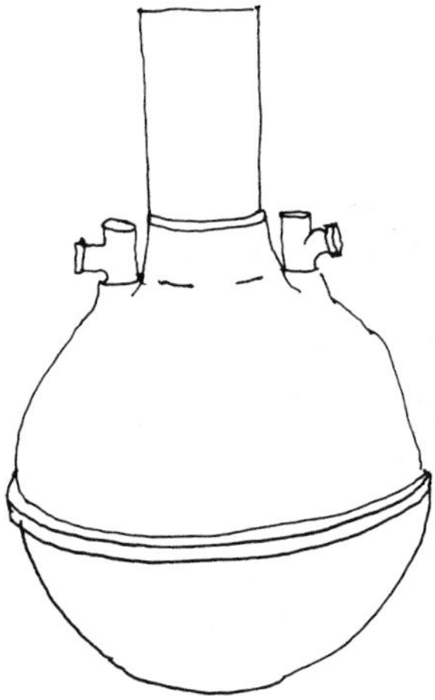

Interceptor

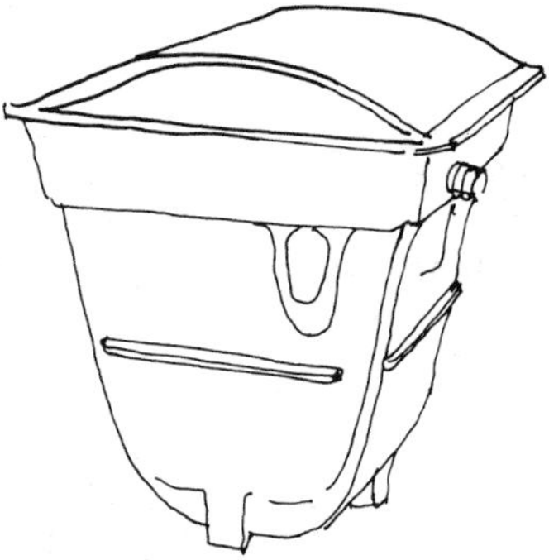

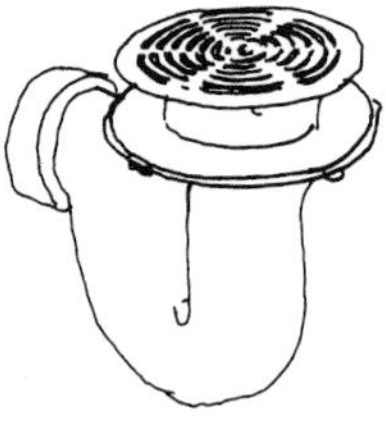

Floor drain

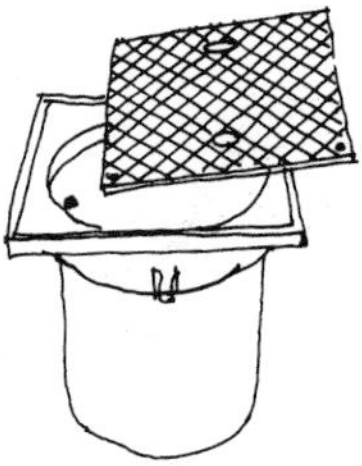

Rodding eye

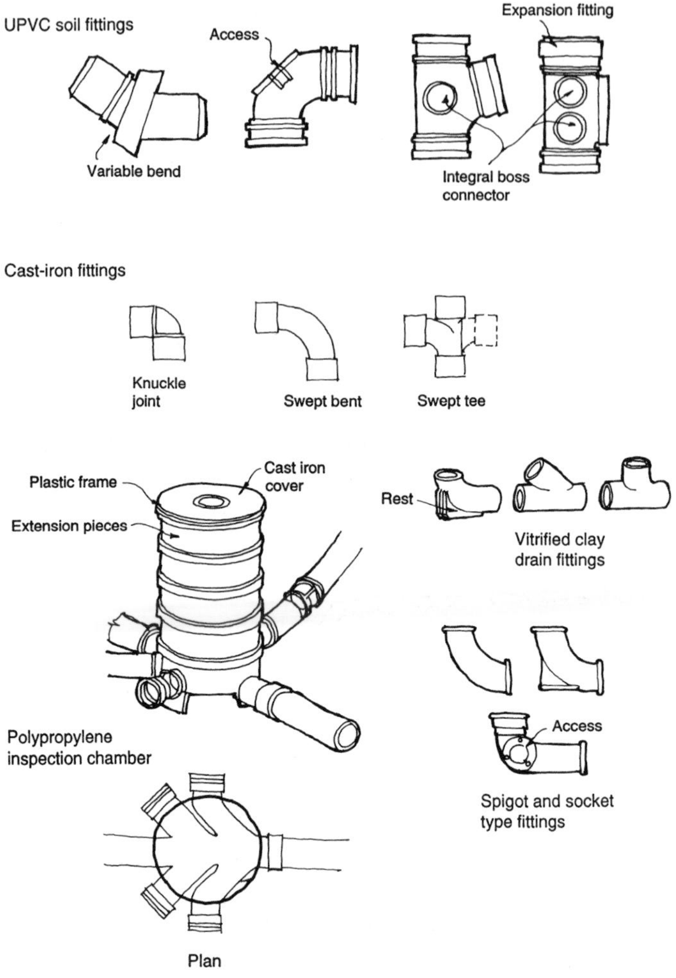
UPVC soil fittings
Access
Expansion fitting
Variable bend
Integral boss connector
Cast-iron fittings
Knuckle joint
Swept bent
Swept tee
Plastic frame
Cast iron cover
Extension pieces
Rest
Vitrified clay drain fittings
Polypropylene inspection chamber
Access
Spigot and socket type fittings
Plan

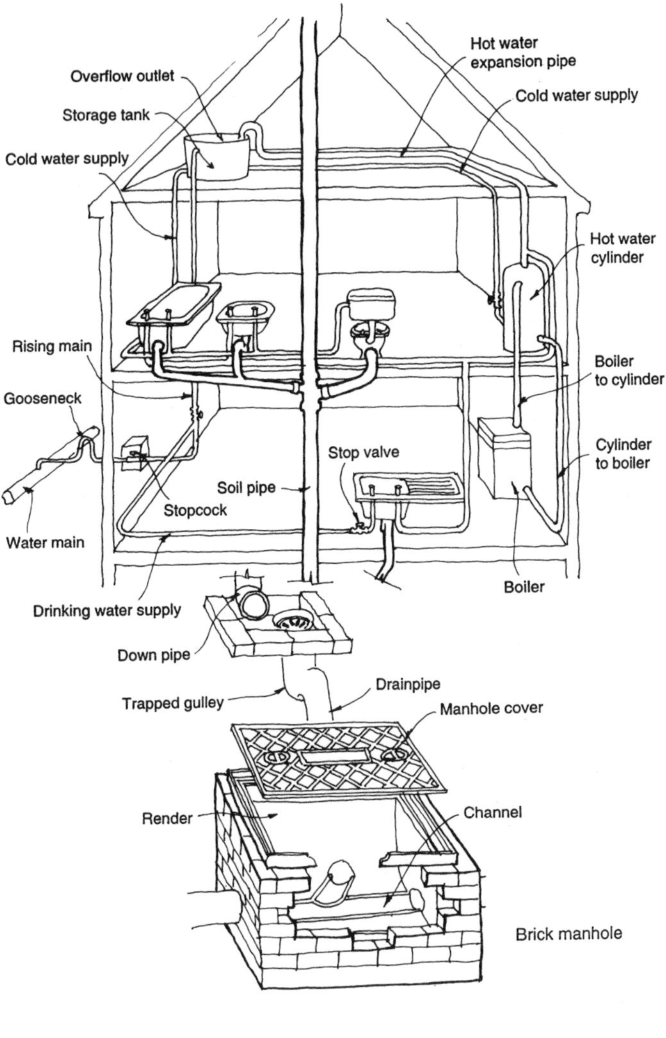
Hot water
expansion pipe
Overflow outlet
Cold water supply
Storage tank
Cold water supply
Hot water
cylinder
Rising main
Boiler
to cylinder
Gooseneck
Stop valve
Cylinder
to boiler
Soil pipe
Stopcock
Water main
Boiler
Drinking water supply
Down pipe
Drainpipe
Trapped gulley
Manhole cover
Render
Channel
Brick manhole

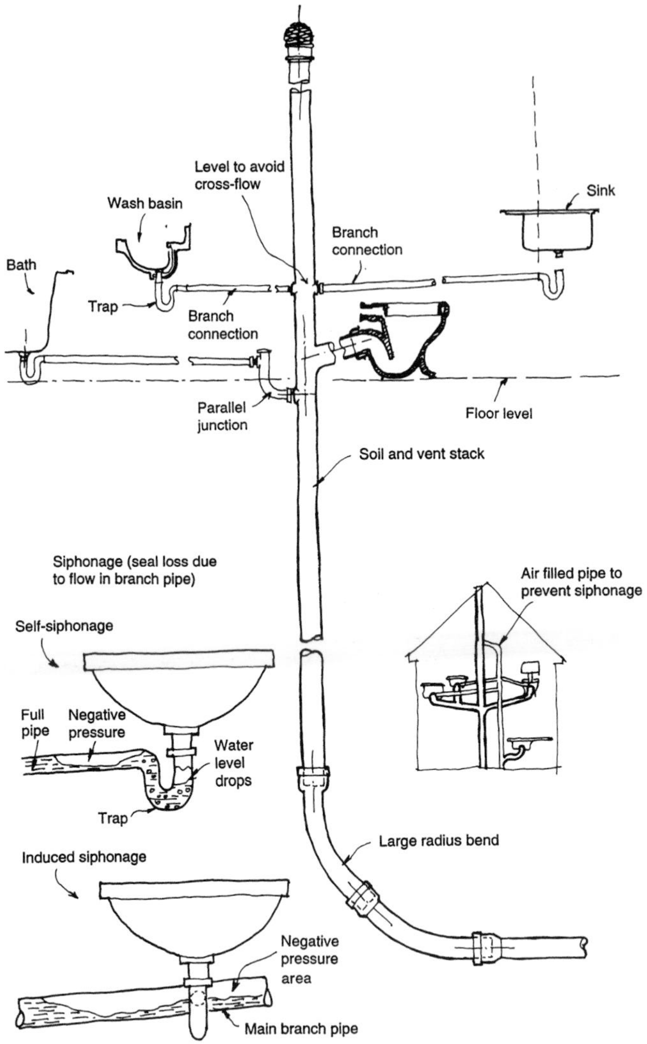
Level to avoid cross-flow
Wash basin
Sink
Branch connection
Bath
Trap
Branch connection
Parallel junction
Floor level
Soil and vent stack
Siphonage (seal loss due to flow in branch pipe)
Air filled pipe to prevent siphonage
Self-siphonage
Full pipe
Negative pressure
Water level drops
Trap
Large radius bend
Induced siphonage
Negative pressure area
Main branch pipe

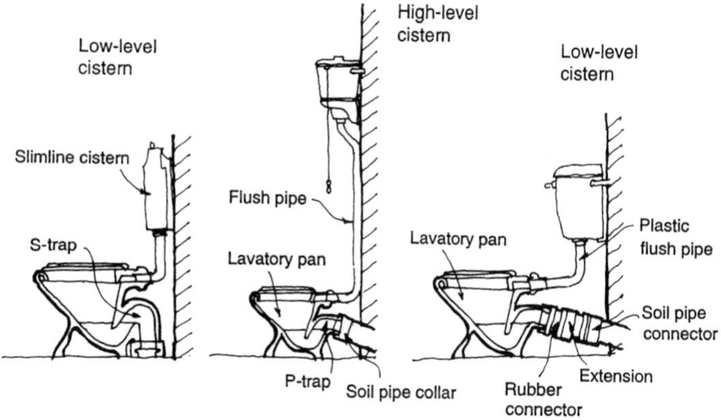
Low-level
cistern
Slimline cistern
S-trap
High-level
cistern
Flush pipe
Lavatory pan
P-trap
Soil pipe collar
Low-level
cistern
Lavatory pan
Plastic
flush pipe
Soil pipe
connector
Extension
Rubber
connector

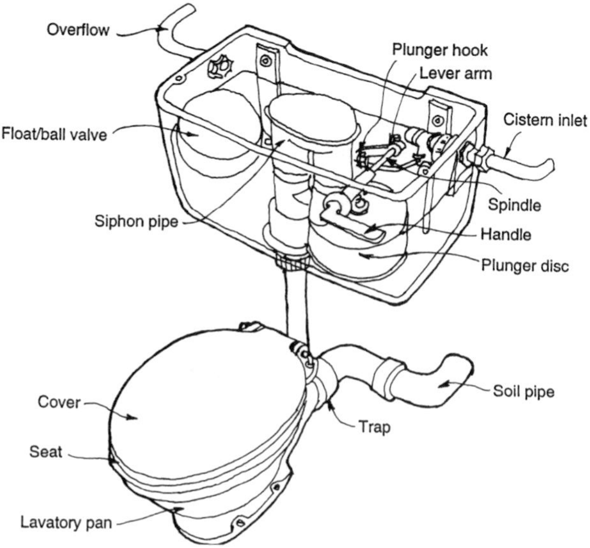
Overflow
Plunger hook
Lever arm
Cistern inlet
Float/ball valve
Spindle
Siphon pipe
Handle
Plunger disc
Cover
Soil pipe
Trap
Seat
Lavatory pan

Cold water tank

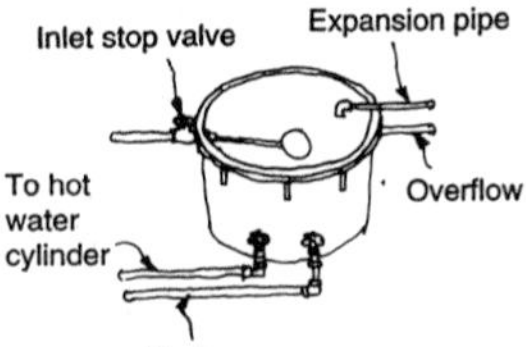

Portsmouth valve

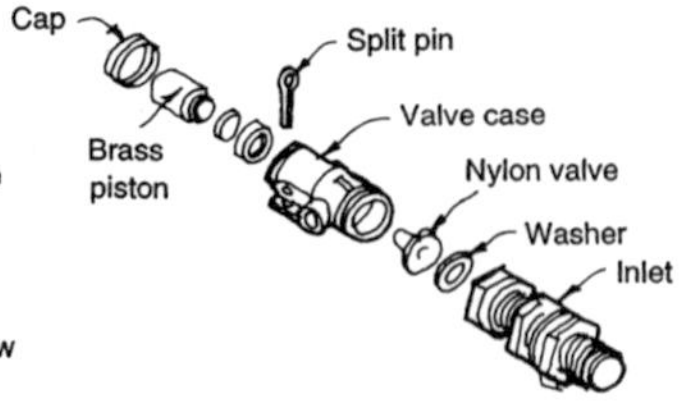

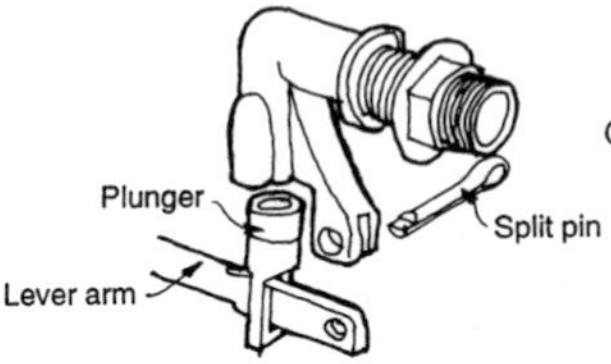

Croydon valve

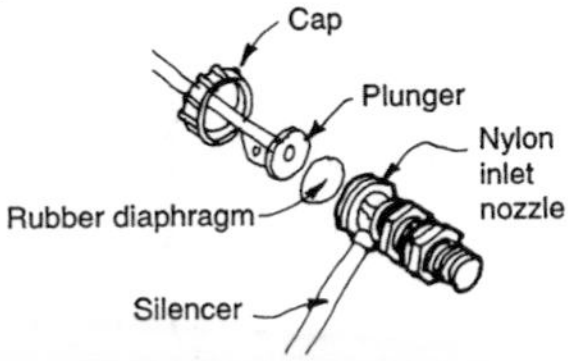

Diaphragm valve

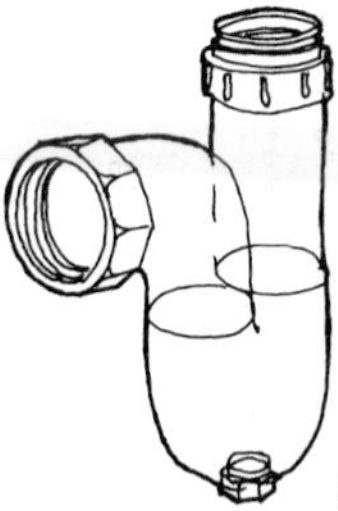

U-trap

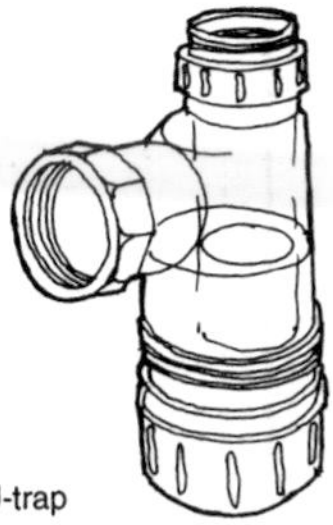

Bottle trap

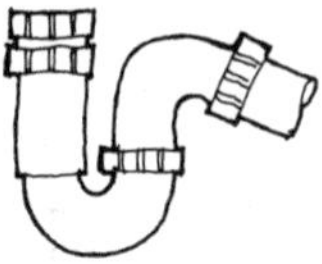

P-trap outlet

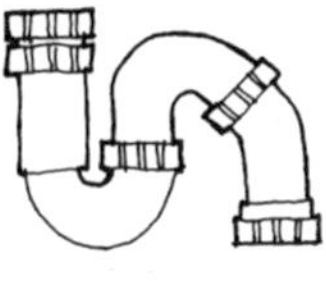

S-trap outlet

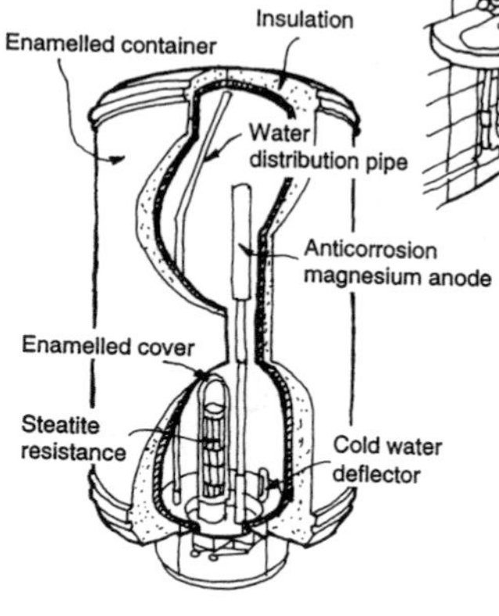

Electric storage heater

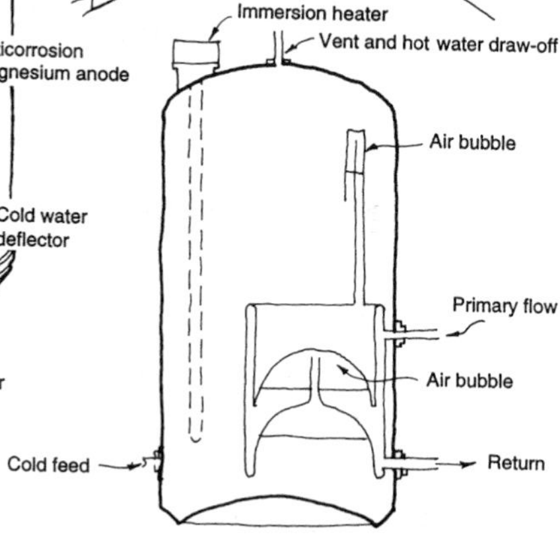

Patent indirect hot water cylinder

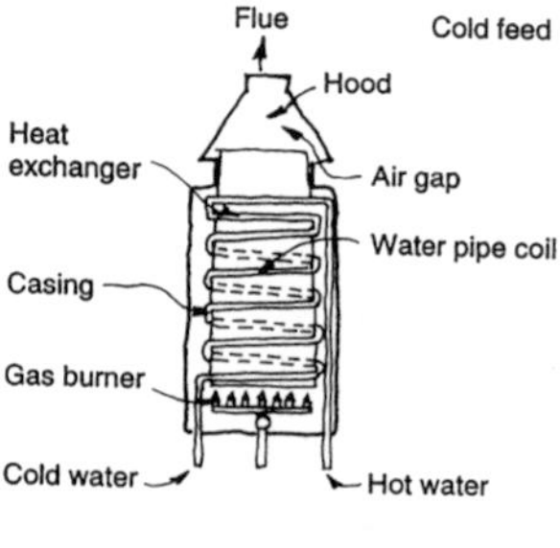

Gas 'instant' water heater

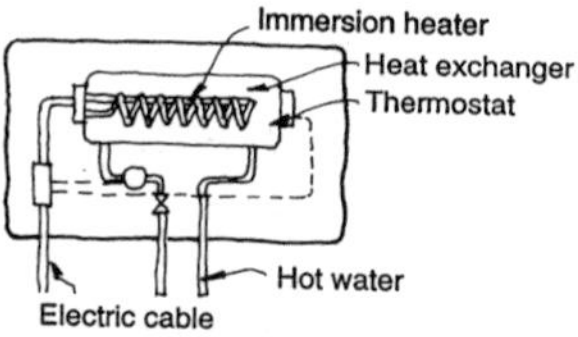

Electric 'instant' heater

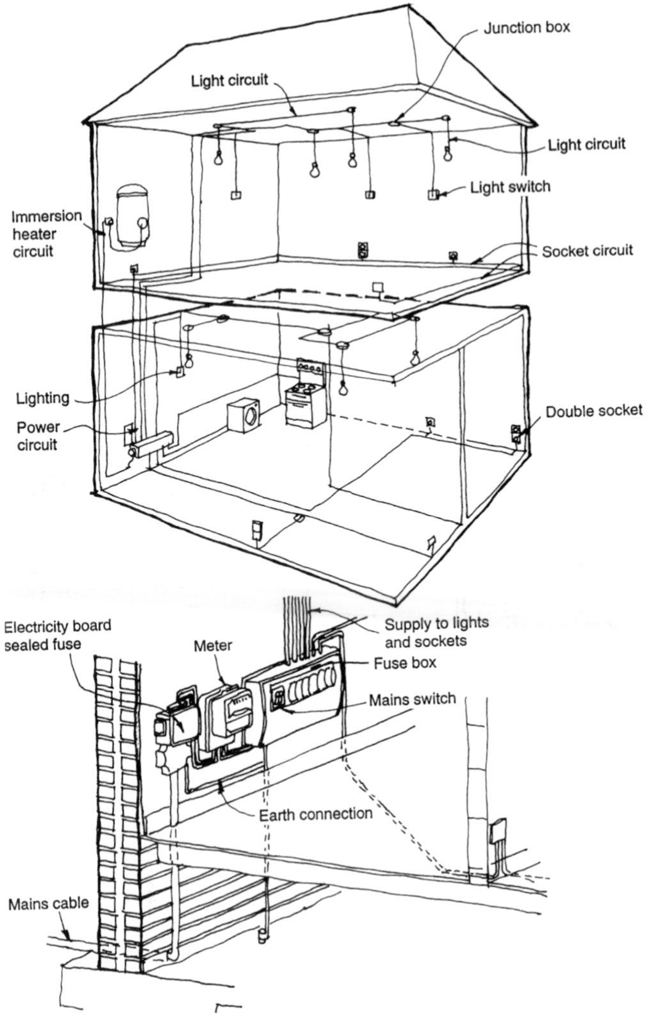
Junction box
Light circuit
Light circuit
Light switch
Immersion heater circuit
Socket circuit
Lighting
Double socket
Power circuit
Electricity board sealed fuse
Meter
Supply to lights and sockets
Fuse box
Mains switch
Earth connection
Mains cable

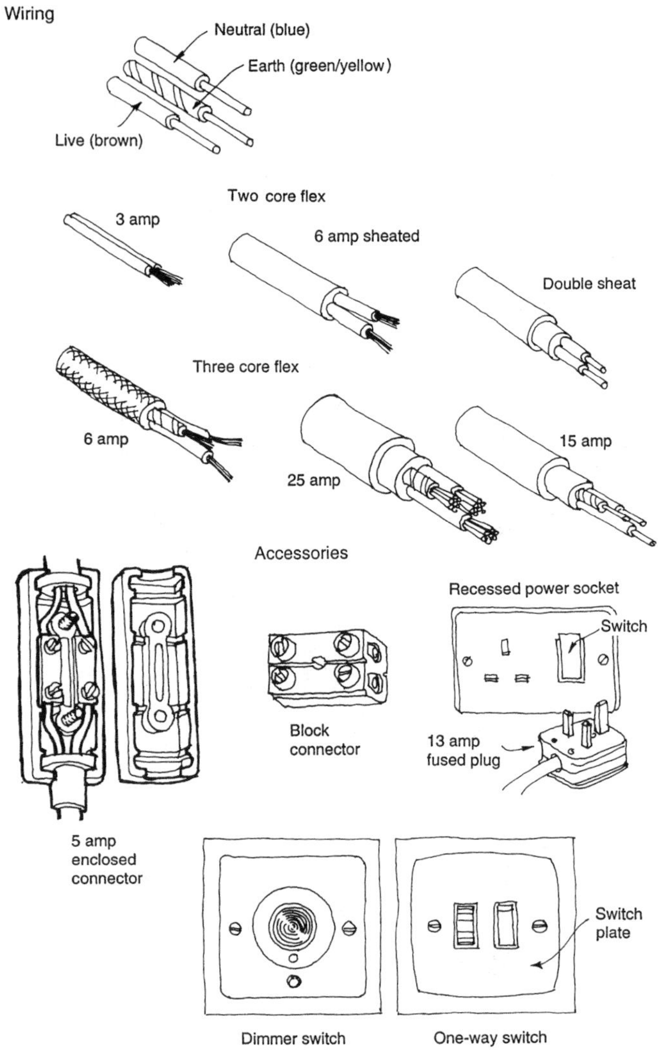
Wiring
Neutral (blue)
Earth (green/yellow)
Live (brown)
Two core flex
3 amp
6 amp sheated
Double sheat
Three core flex
6 amp
25 amp
15 amp
Accessories
Recessed power socket
Switch
Block connector
13 amp fused plug
5 amp enclosed connector
Switch plate
Dimmer switch
One-way switch

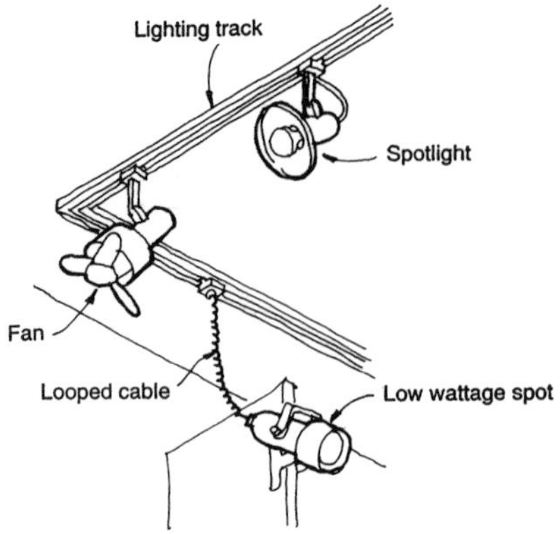
Lighting track
Spotlight
Fan
Looped cable
Low wattage spot

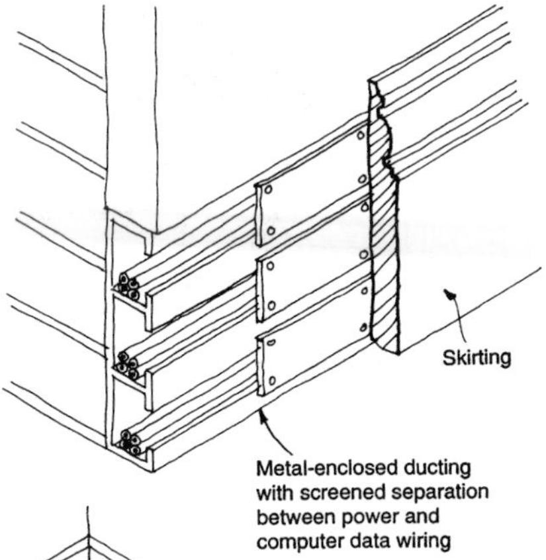
Skirting
Metal-enclosed ducting with screened separation between power and computer data wiring

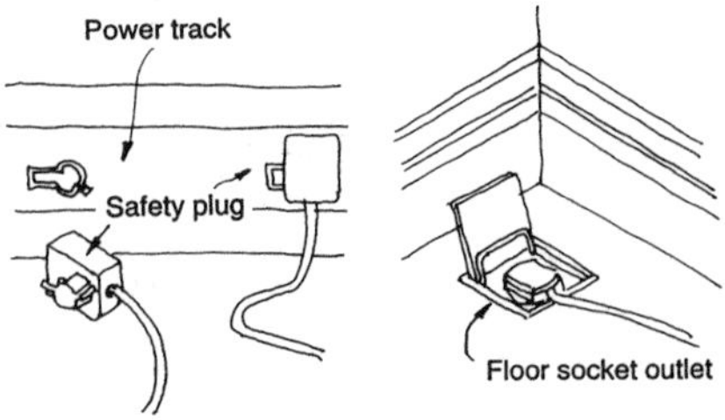
Power track
Safety plug
Floor socket outlet

Outdoor light fittings for plants, paths

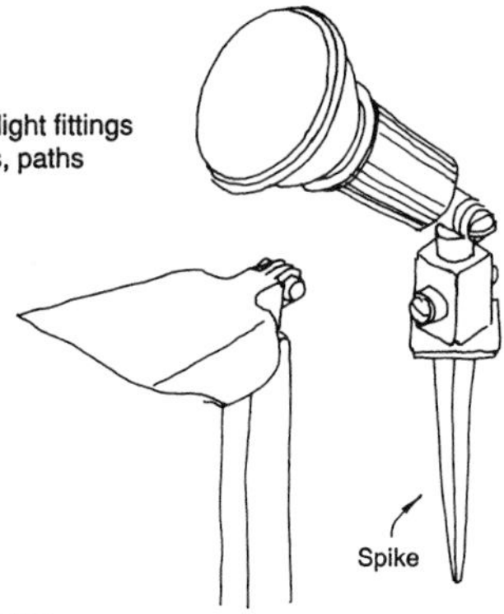

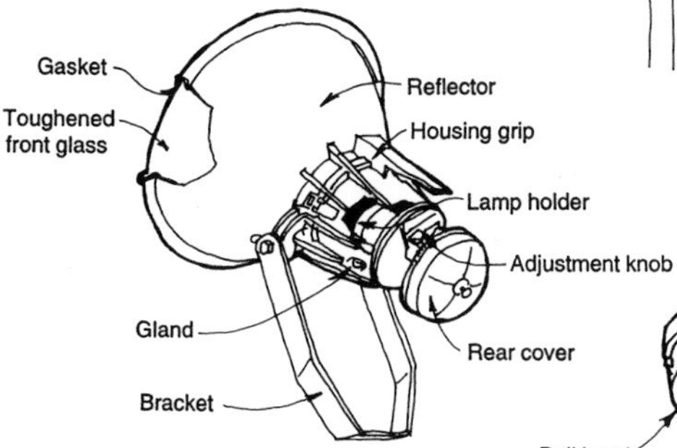

Floodlight projector

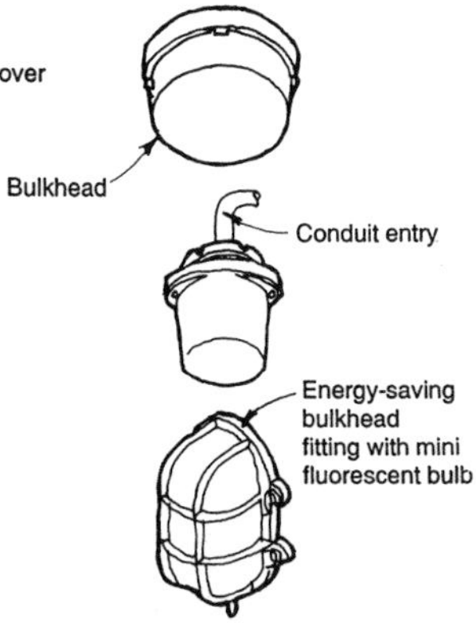

Outdoor lighting fittings

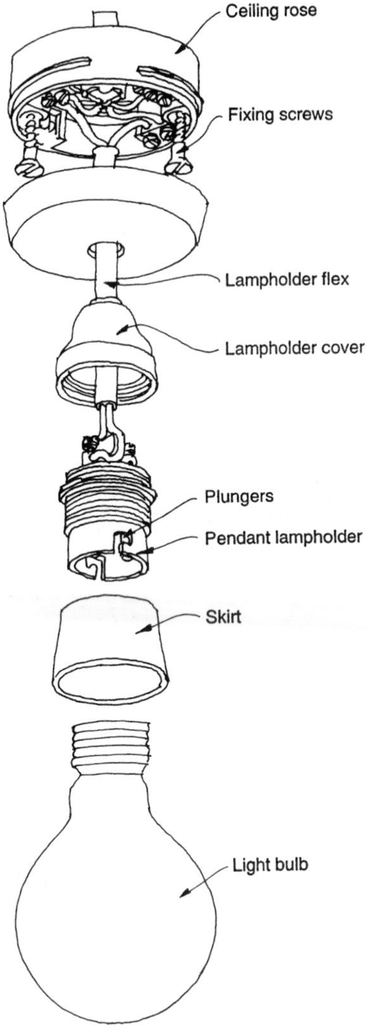
Ceiling rose
Fixing screws
Lampholder flex
Lampholder cover
Plungers
Pendant lampholder
Skirt
Light bulb

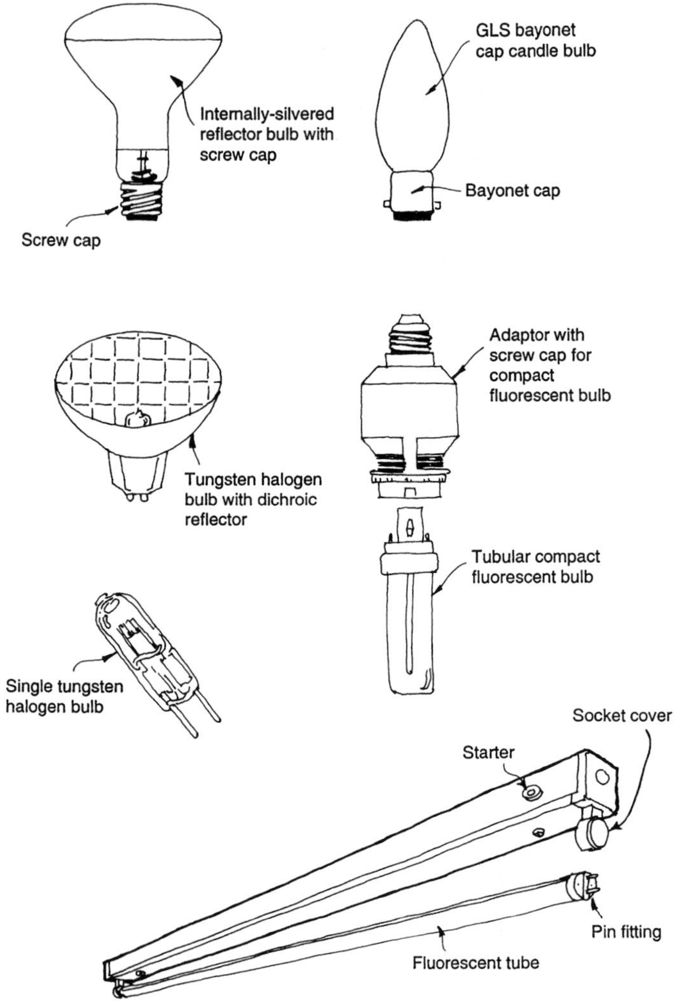
Internally-silvered reflector bulb with screw cap
Screw cap
GLS bayonet cap candle bulb
Bayonet cap
Adaptor with screw cap for compact fluorescent bulb
Tungsten halogen bulb with dichroic reflector
Tubular compact fluorescent bulb
Single tungsten halogen bulb
Socket cover
Starter
Pin fitting
Fluorescent tube

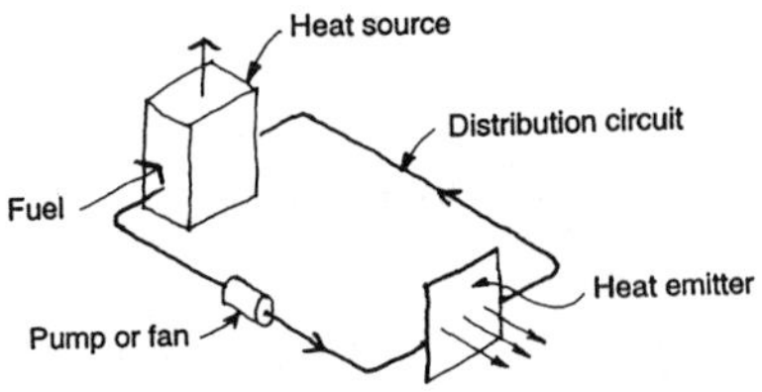
Heat source
Distribution circuit
Fuel
Heat emitter
Pump or fan

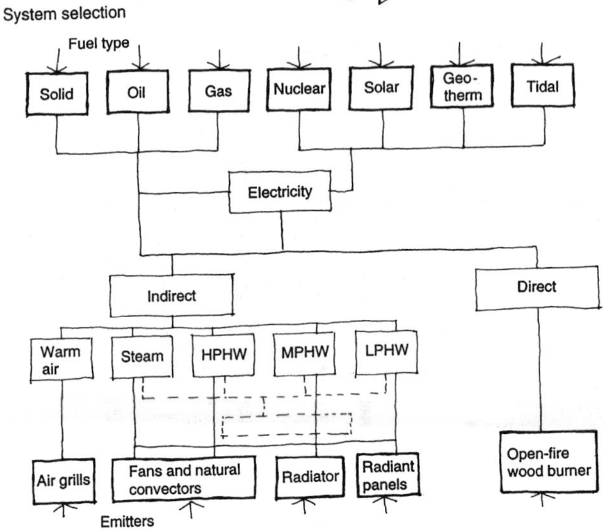
System selection
Fuel type
Solid
Oil
Gas
Nuclear
Solar
Geo-
therm
Tidal
Electricity
Indirect
Direct
Warm
air
Steam
HPHW
MPHW
LPHW
Air grills
Fans and natural
convectors
Radiator
Radiant
panels
Open-fire
wood burner
Emitters

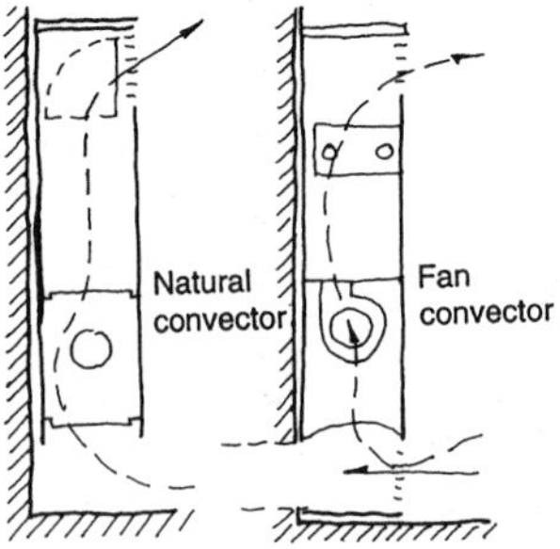
Natural
convector
Fan
convector

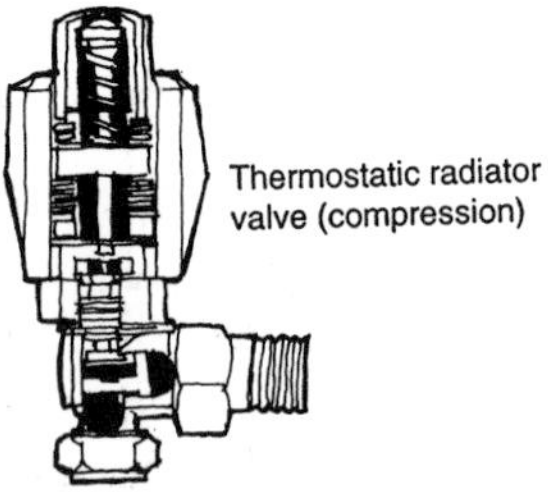
Thermostatic radiator
valve (compression)

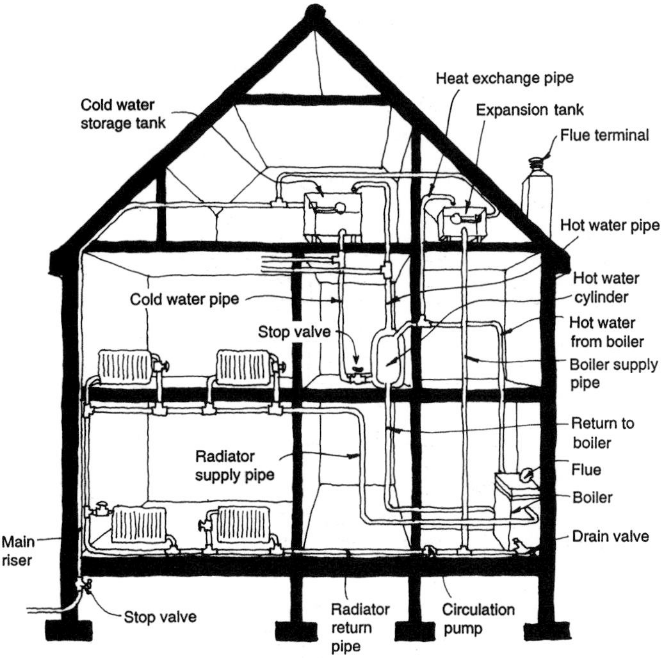

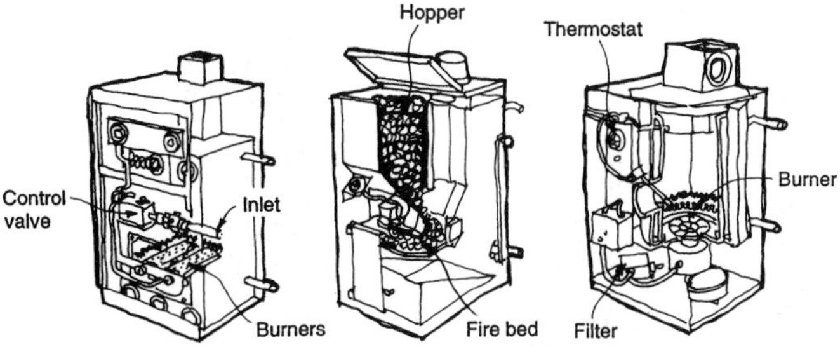

Gas-fired boiler

Solid-fuel boiler

Oil-fired boiler

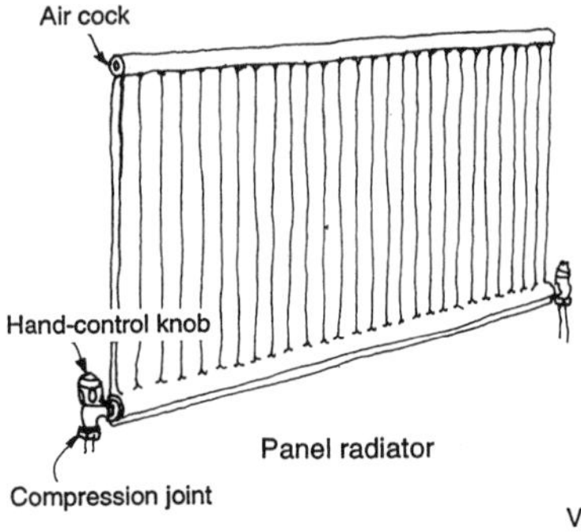

Panel radiator

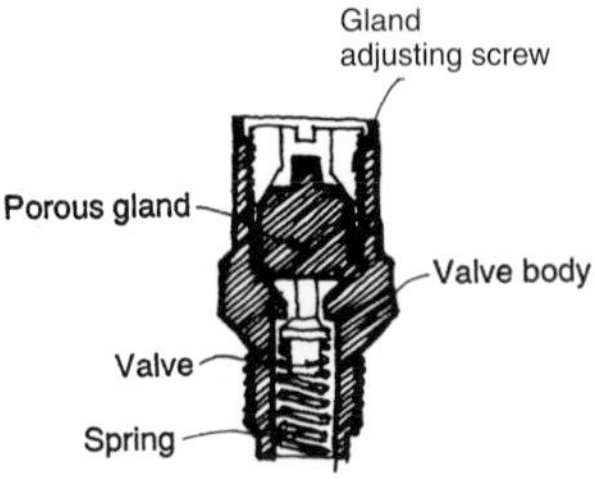

Automatic
air eliminator

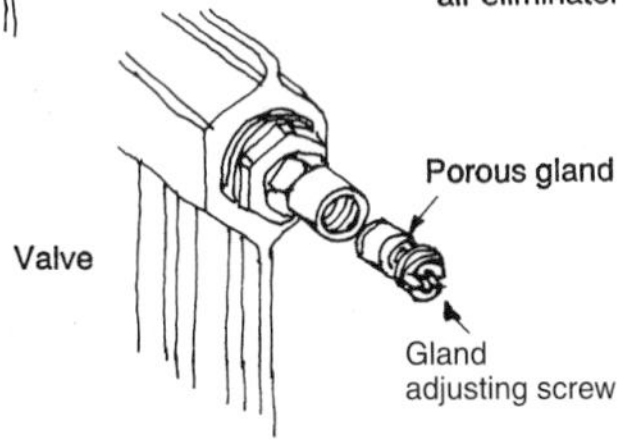

Water-filled

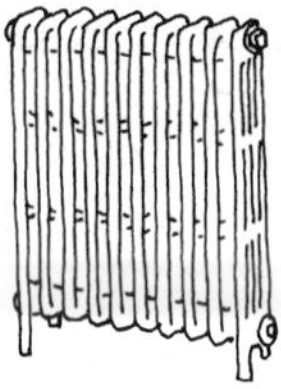

Column radiator

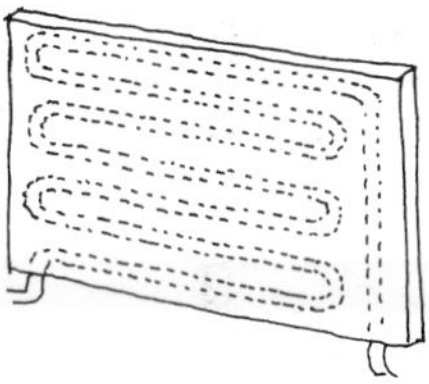

Electrical
radiant panel

Storage heater

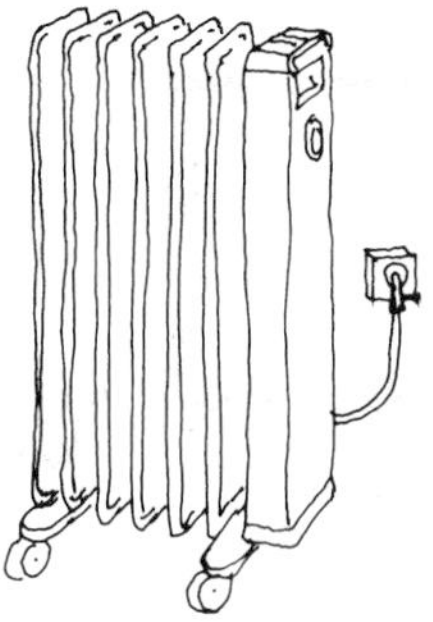

Oil-filled radiator

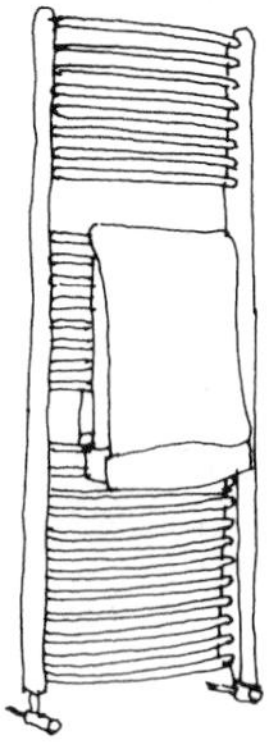

Tower heater

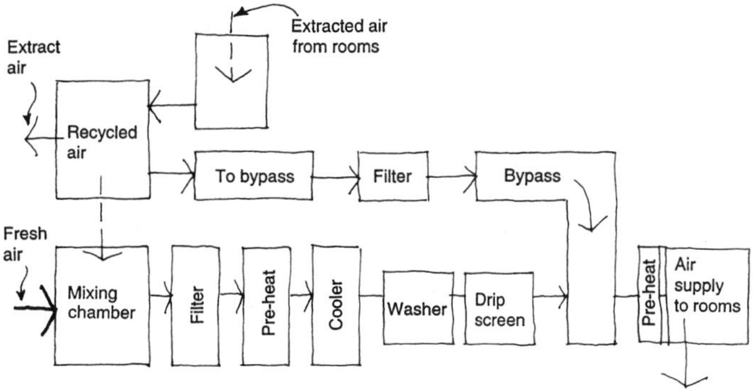

Air conditioning plant sequence

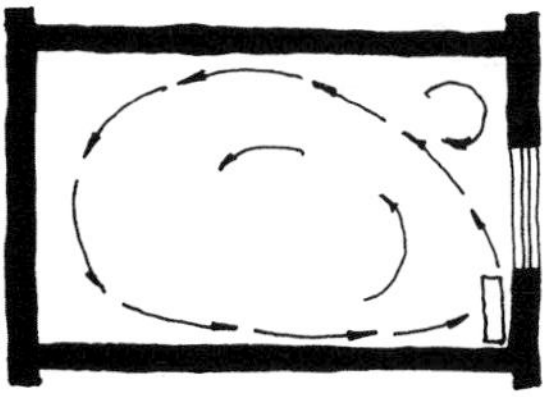

Unventilated room

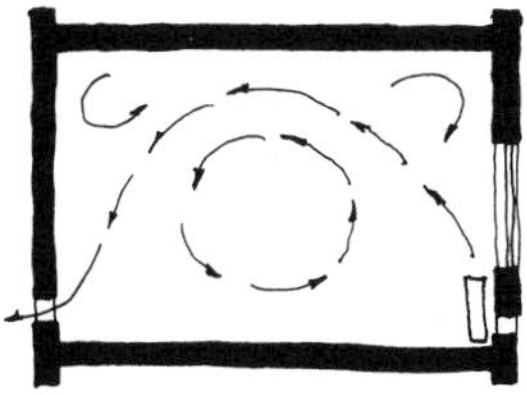

Room with
pre-heated fresh air

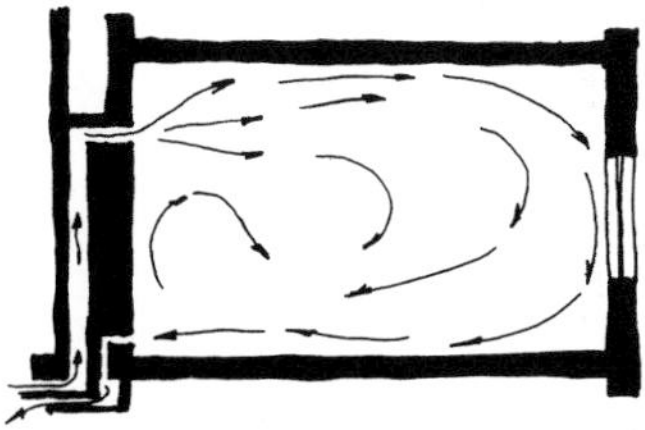

High supply, low extract

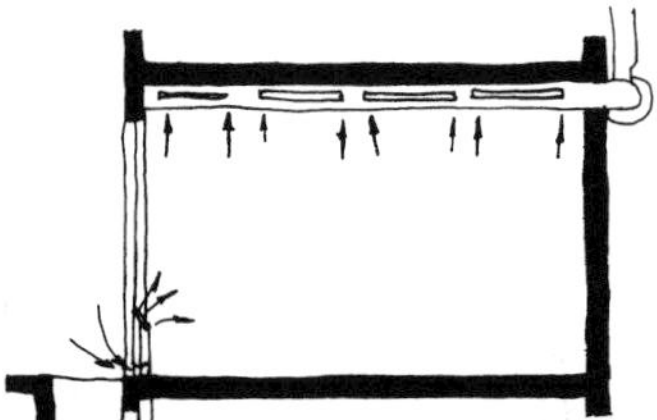

Improved ventilation with
slots to full width

Graphic presentation of earthwork

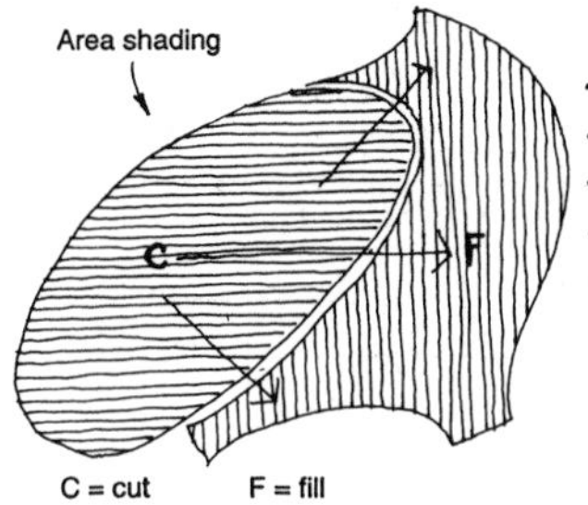
Area shading
C
F
C = cut
F = fill

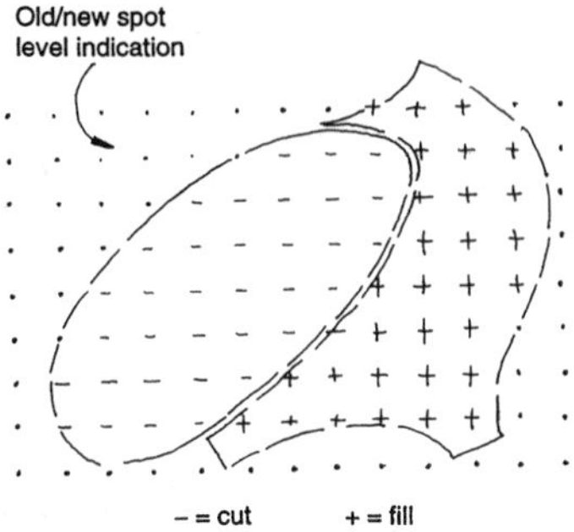
Old/new spot level indication
– = cut
+ = fill

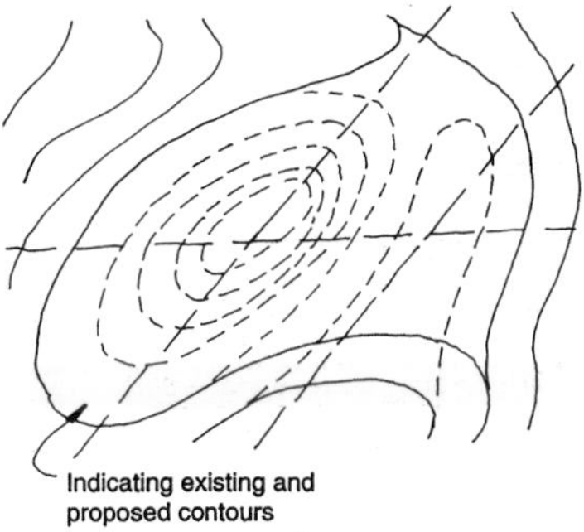
Indicating existing and proposed contours

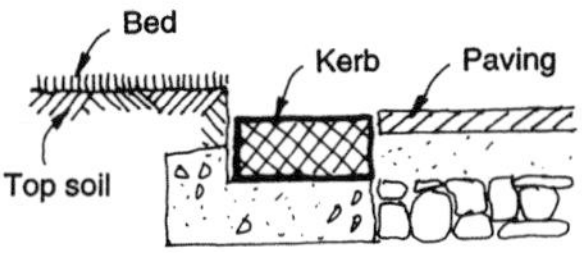
Bed
Kerb
Paving
Top soil

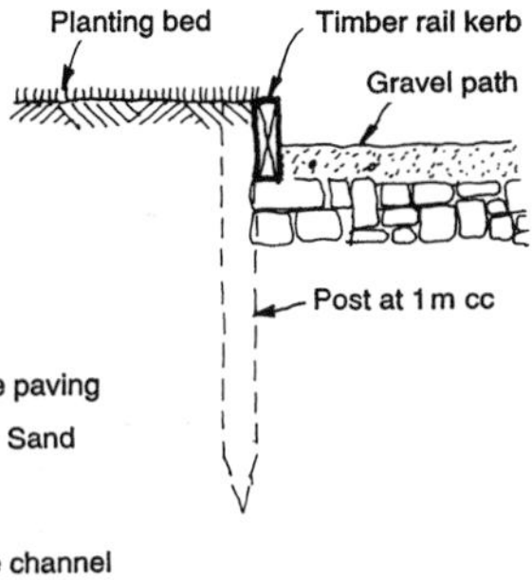
Planting bed
Timber rail kerb
Gravel path
Post at 1 m cc

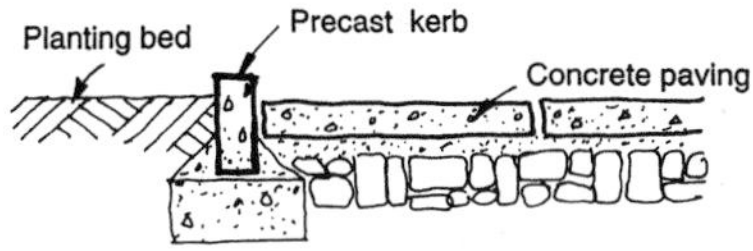
Planting bed
Precast kerb
Concrete paving

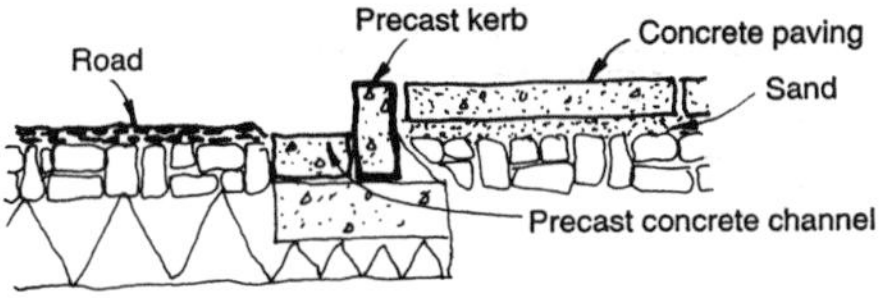
Road
Precast kerb
Concrete paving
Sand
Precast concrete channel

Irrigation system

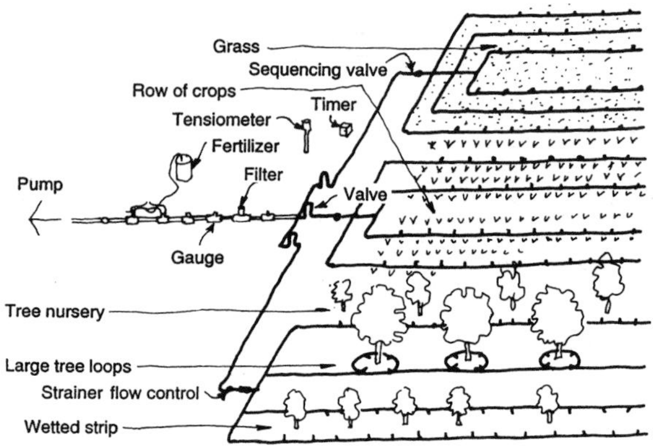
Grass
Sequencing valve
Row of crops
Timer
Tensiometer
Fertilizer
Filter
Pump
Valve
Gauge
Tree nursery
Large tree loops
Strainer flow control
Wetted strip

Screening

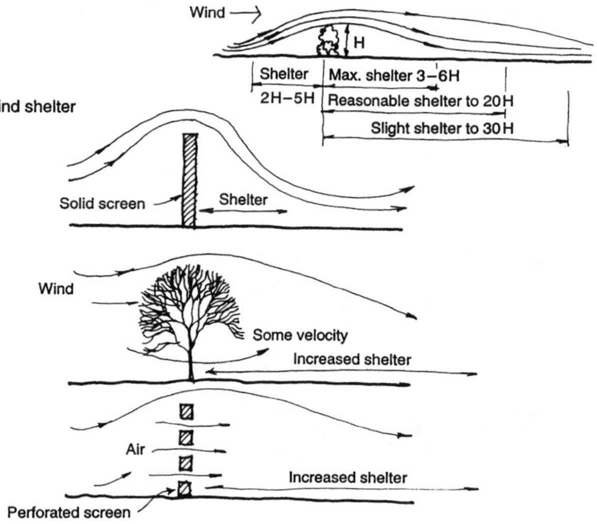
Wind
H
Shelter
2H–5H
Max. shelter 3–6H
Reasonable shelter to 20H
Slight shelter to 30H
Wind shelter
Solid screen
Shelter
Wind
Some velocity
Increased shelter
Air
Increased shelter
Perforated screen

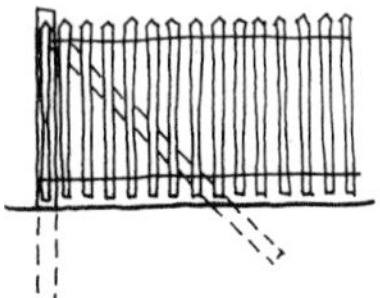

Cleft chestnut paling

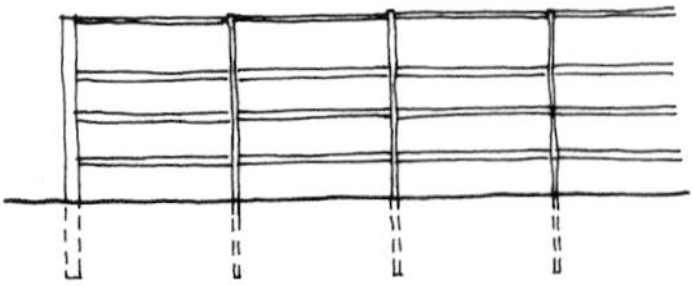

Continuous bar fencing

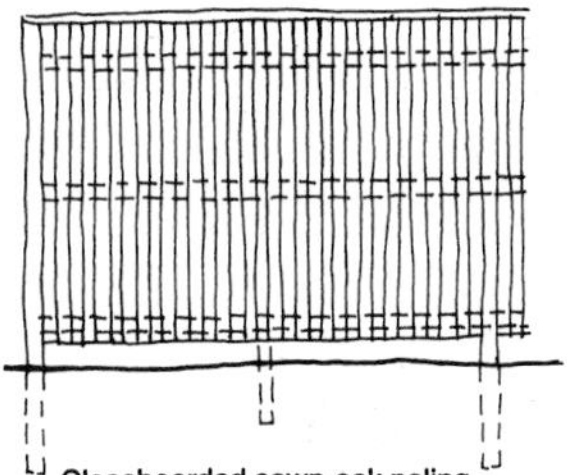

Closeboarded sawn-oak paling

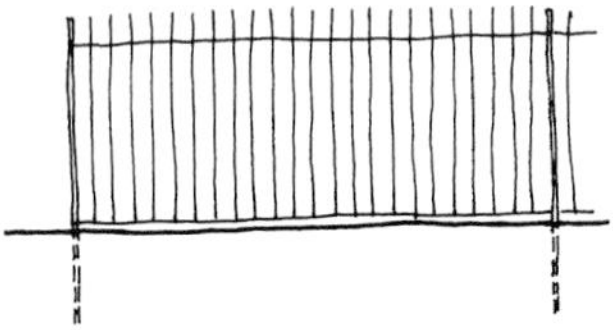

Vertical bar fencing

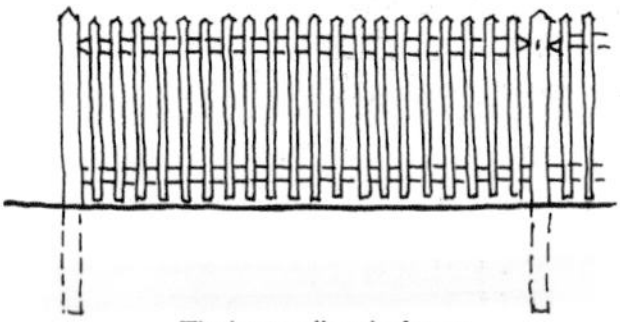

Timber palisade fence

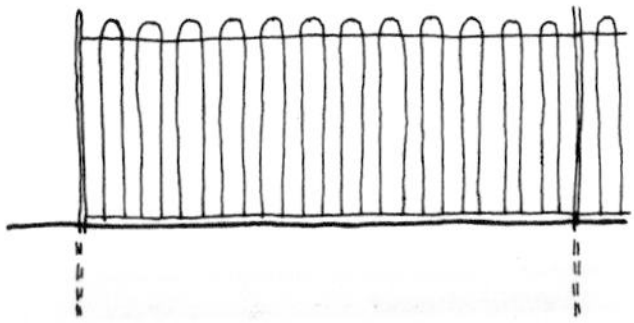

Vertical bar fencing with hair pin tops

Woven wood fence

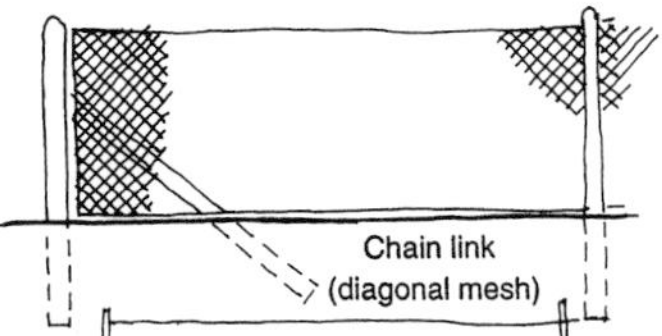

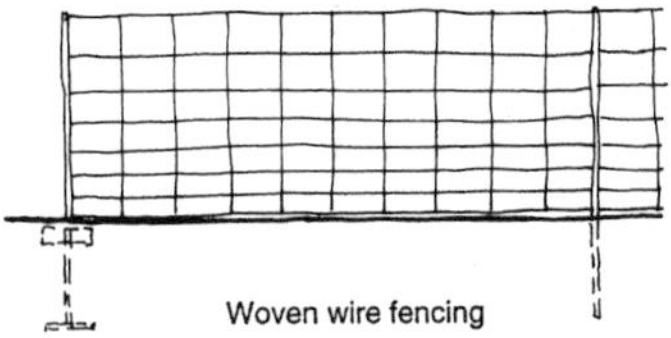

Woven wire fencing

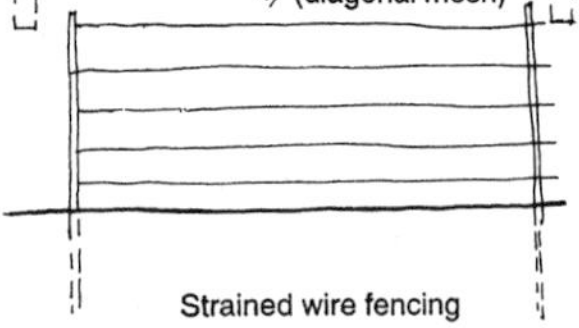

Strained wire fencing

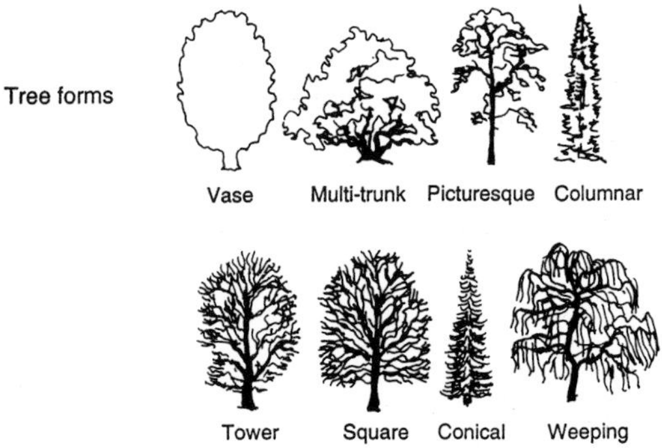
Tree forms
Vase
Multi-trunk
Picturesque
Columnar
Tower
Square
Conical
Weeping

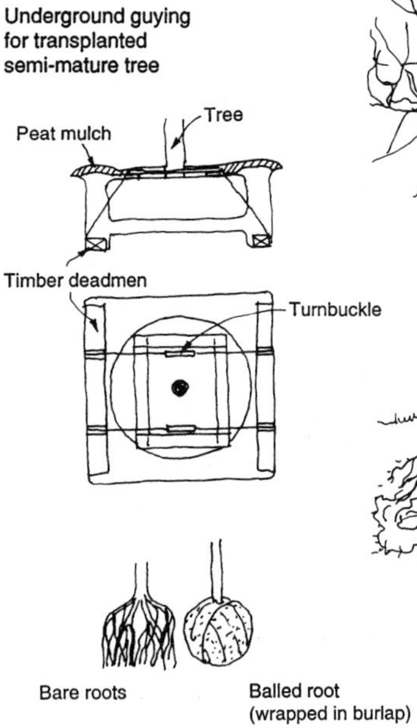
Underground guying
for transplanted
semi-mature tree
Tree
Peat mulch
Timber deadmen
Turnbuckle
Bare roots
Balled root
(wrapped in burlap)

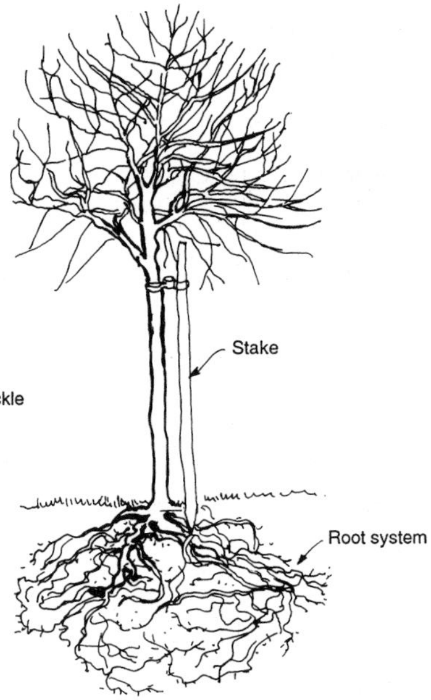
Stake
Root system

Grading

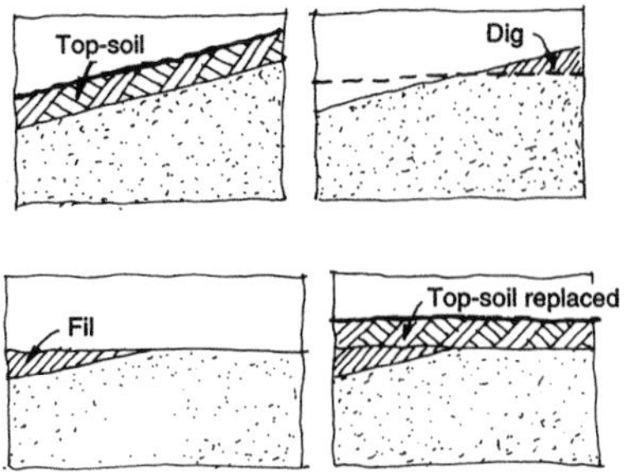

Land drainage systems

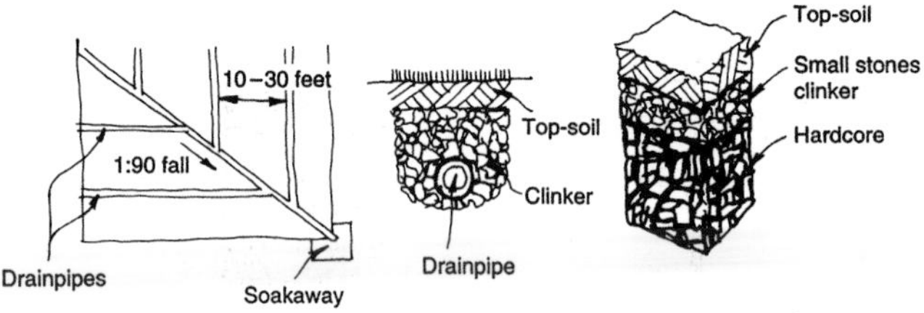

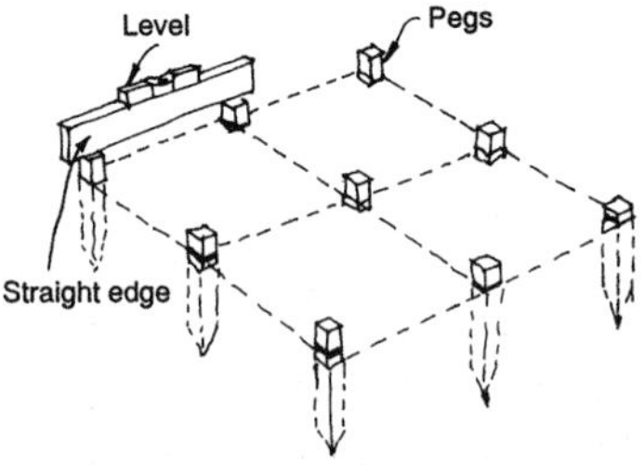

Levelling method

Turf laying

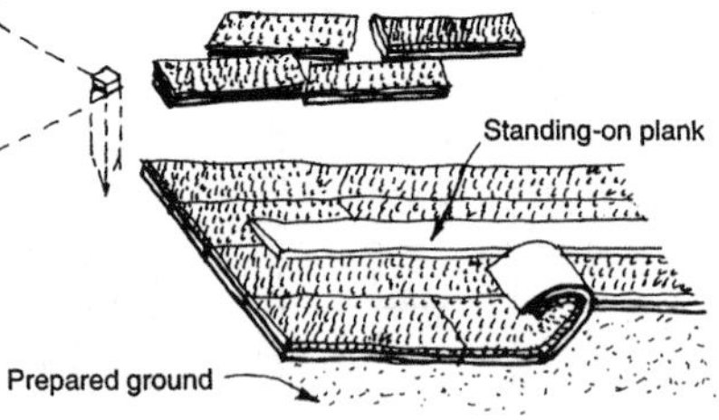

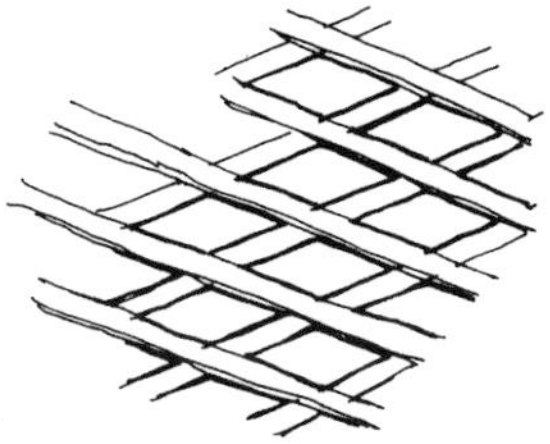
Trellis

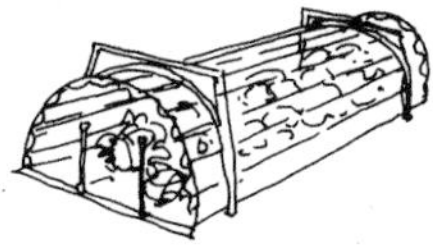
Cloche

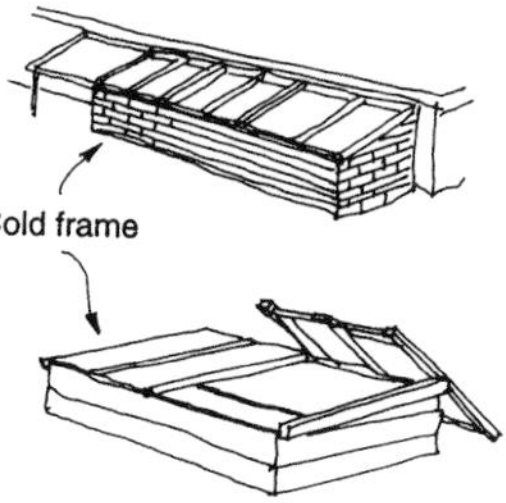
Cold frame

Greenhouse staging

Greenhouse

Gazebo

Conservatory

VI. The Environmental Conditions

GLOBAL WARMING

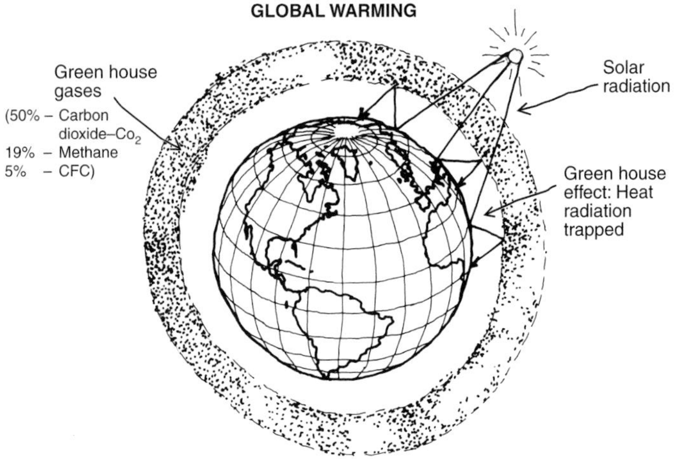
Green house
gases
(50% – Carbon
dioxide–Co2
19% – Methane
5% – CFC)
Solar
radiation
Green house
effect: Heat
radiation
trapped

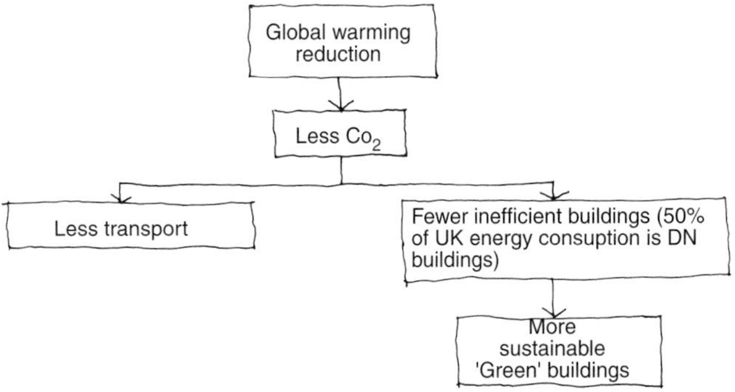
Global warming
reduction
Less Co2
Less transport
Fewer inefficient buildings (50%
of UK energy consuption is DN
buildings)
More
sustainable
'Green' buildings

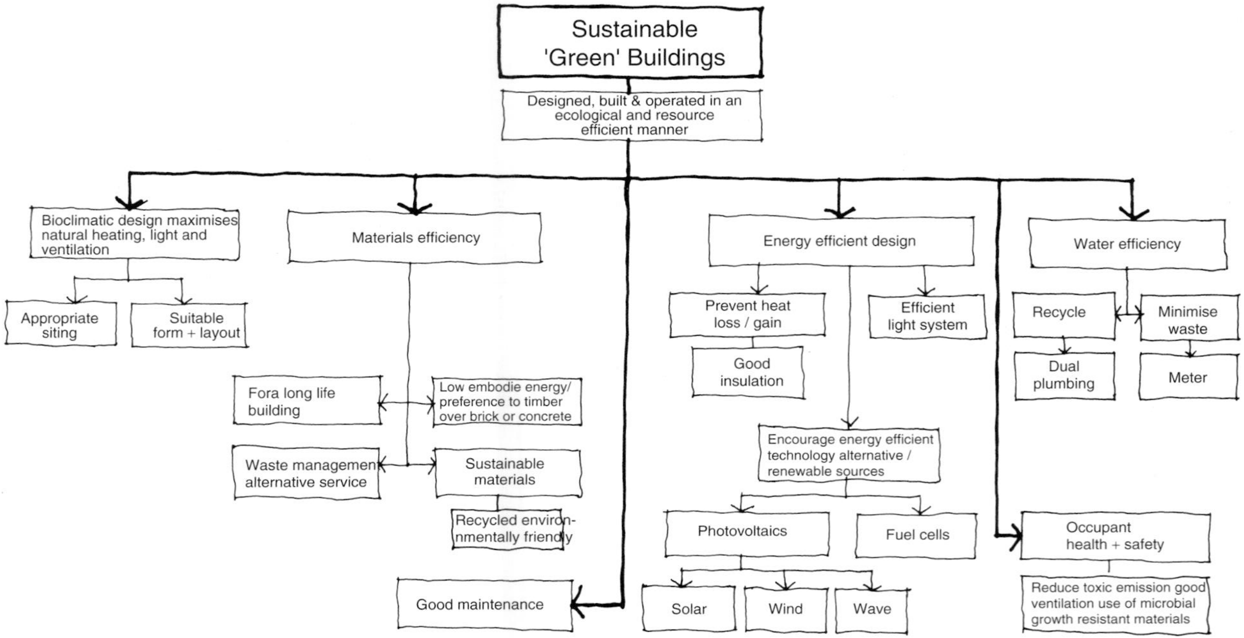

Sustainable 'Green' Buildings
Designed, built & operated in an ecological and resource efficient manner
Bioclimatic design maximises natural heating, light and ventilation
Appropriate siting
Suitable form + layout
Materials efficiency
Fora long life building
Low embodie energy/ preference to timber over brick or concrete
Waste managemen alternative service
Sustainable materials
Recycled environ-nmentally friendly
Good maintenance
Energy efficient design
Prevent heat loss / gain
Good insulation
Efficient light system
Encourage energy efficient technology alternative / renewable sources
Photovoltaics
Fuel cells
Solar
Wind
Wave
Water efficiency
Recycle
Minimise waste
Dual plumbing
Meter
Occupant health + safety
Reduce toxic emission good ventilation use of microbial growth resistant materials

TRADITIONAL BUILDING

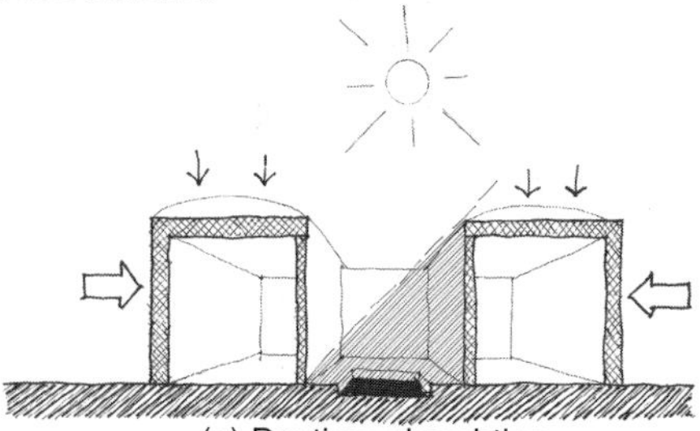

(a) Daytime absorbtion

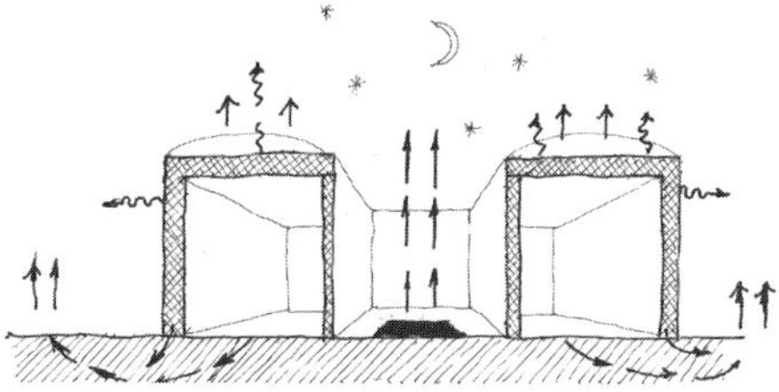

(b) Night time cooling

MODERN BUILDING

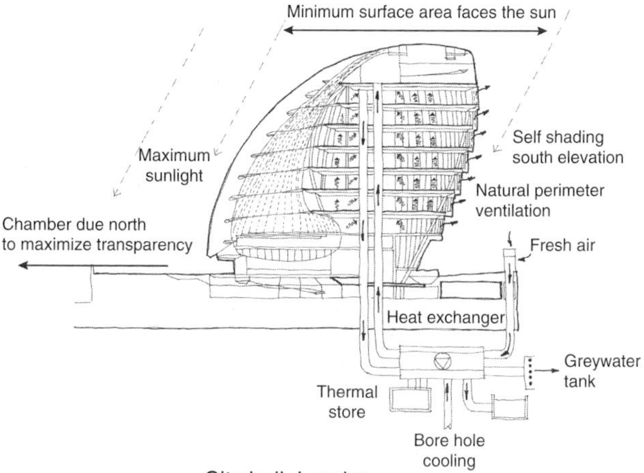

City hall, London

Materials efficiency-waste management

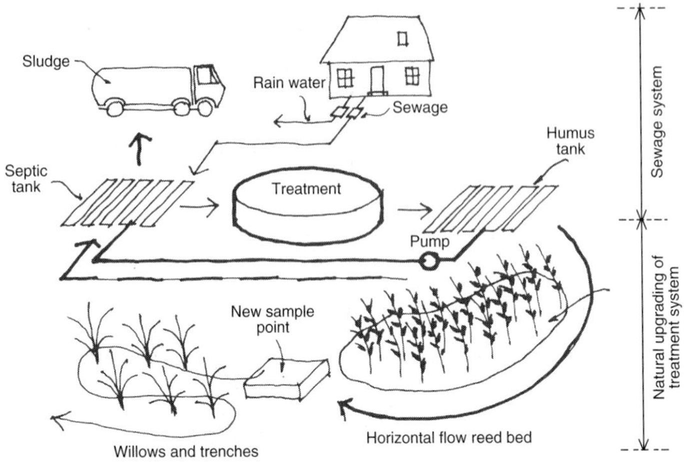

Sewage treatment systems

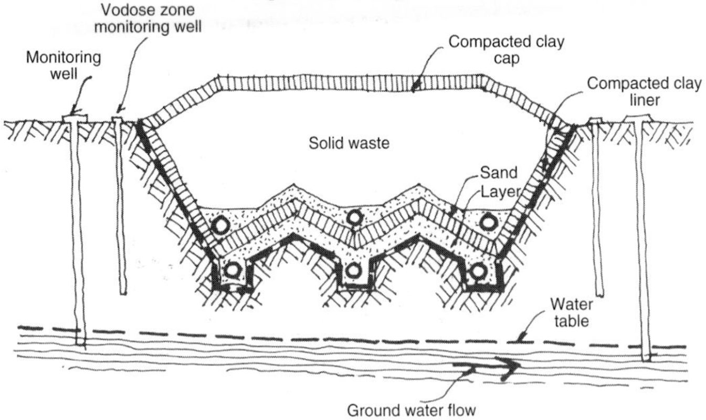

Land fill

Energy efficient design: alternative sources of energy

Solar power

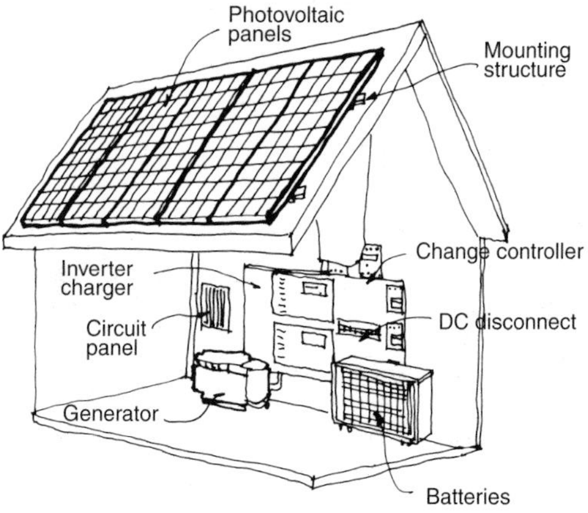

Hybrid power system

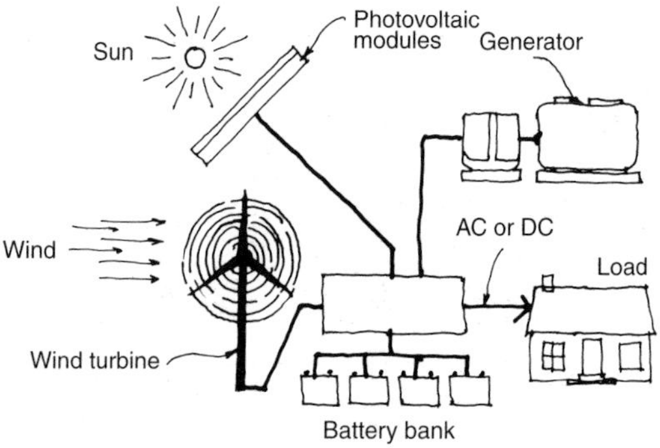

Energy efficient design: alternative sources of energy

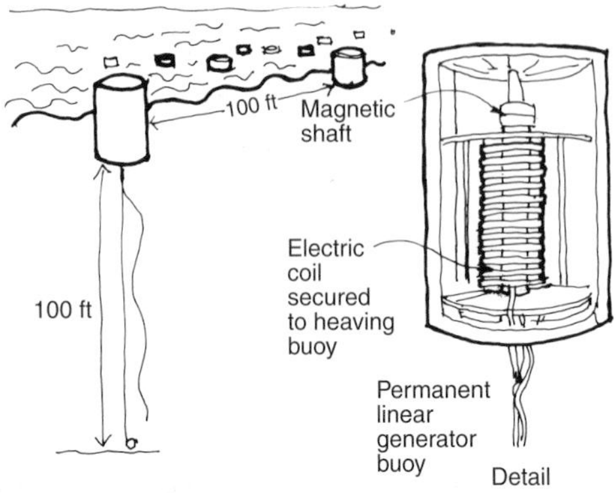

Wave energy

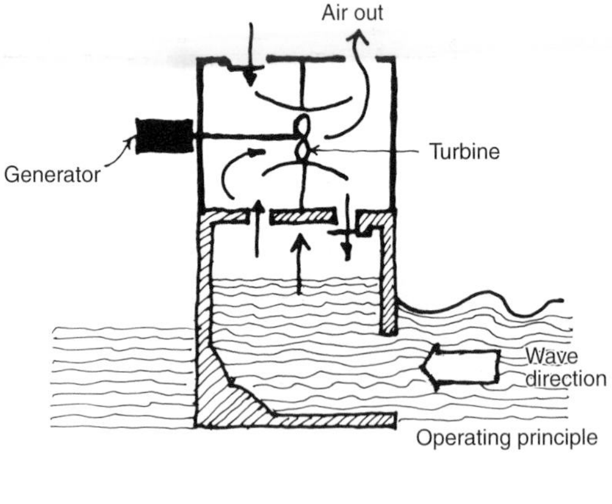

Characteristics

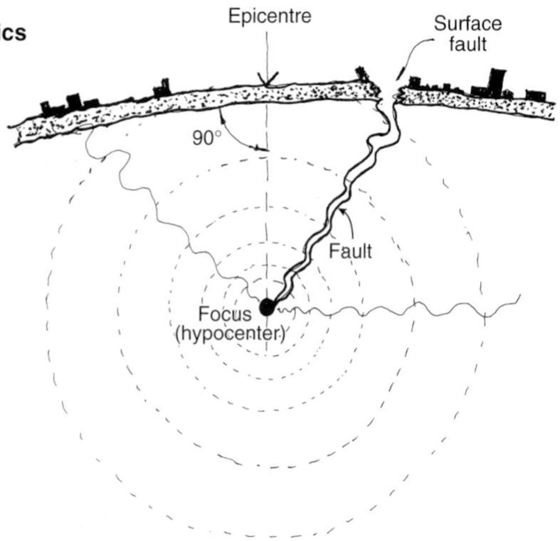

Vibration modes

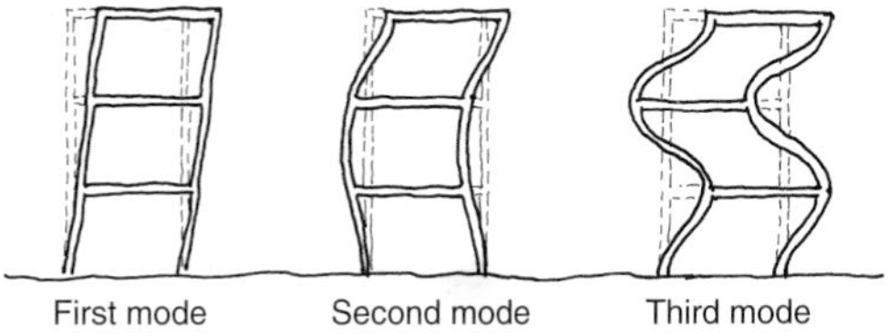

Structural tilting due to liquefaction

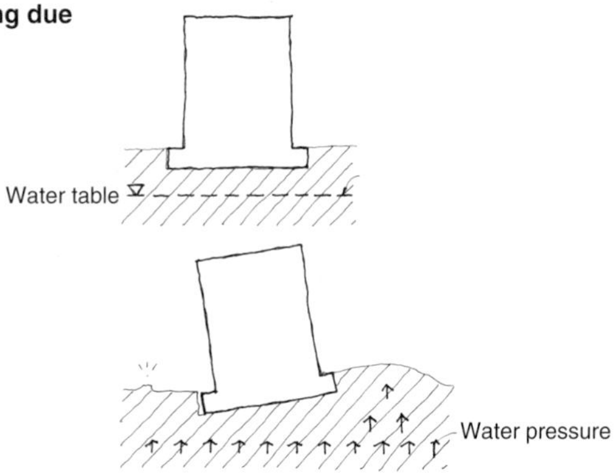

Pancaking

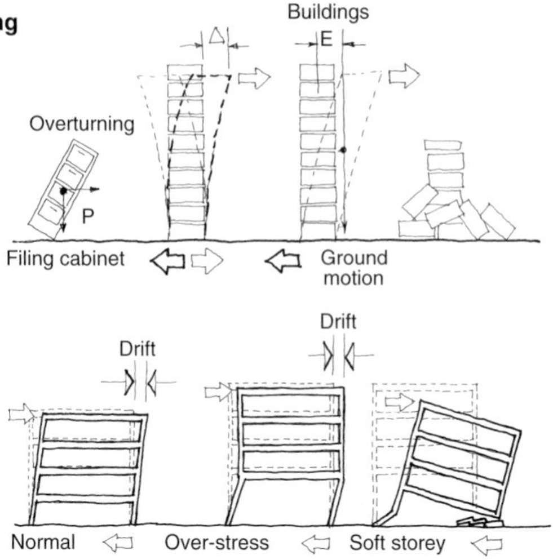
Buildings
Δ
E
Overturning
P
Filing cabinet
Ground
motion
Drift
Drift
Normal
Over-stress
Soft storey

Torsion effect

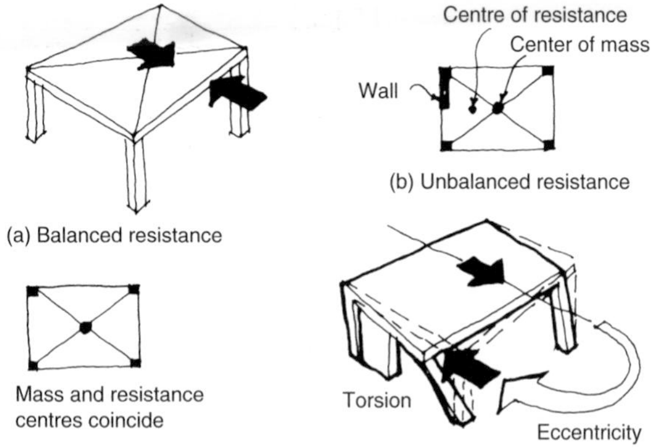
Centre of resistance
Center of mass
Wall
(b) Unbalanced resistance
(a) Balanced resistance
Mass and resistance
centres coincide
Torsion
Eccentricity

Shear failure

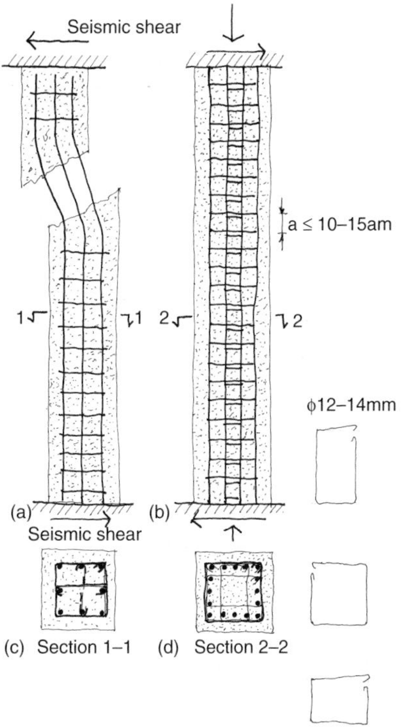
Seismic shear
a ≤ 10–15am
1
1
2
2
ϕ12–14mm
(a)
(b)
Seismic shear
(c) Section 1–1
(d) Section 2–2

Extreme Weather: principles, hurricanes

Hot water circulation principle

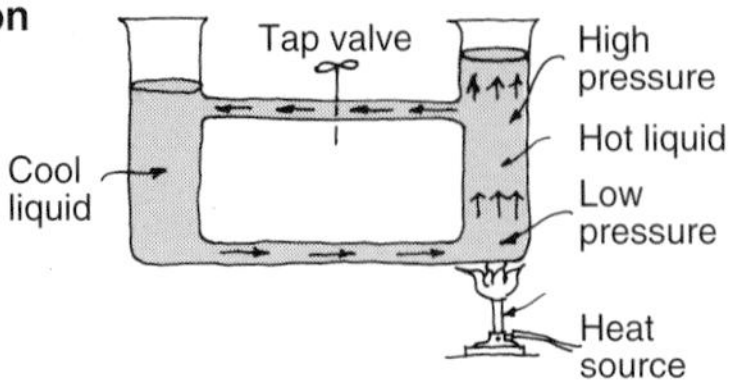

Sea breeze principle

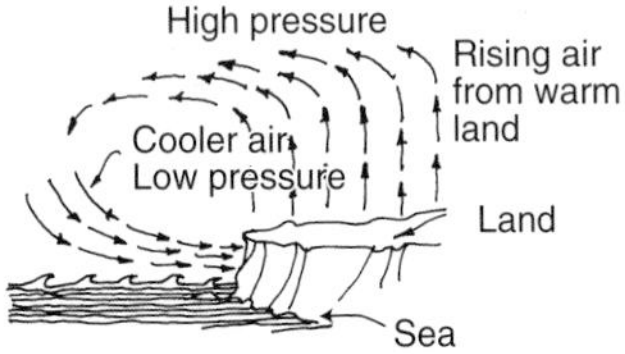

Hurricane development

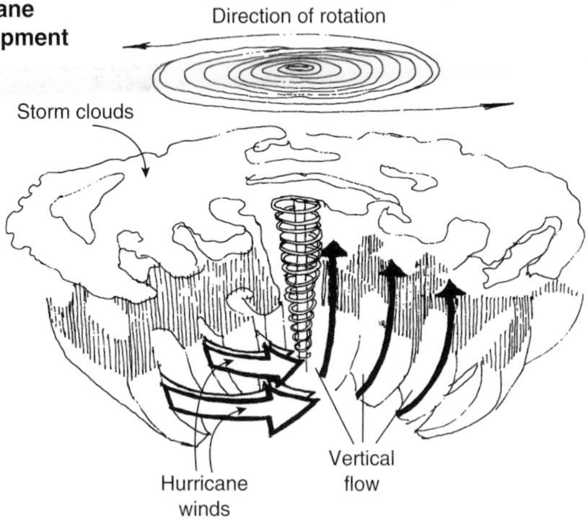

Extreme Weather: hurricanes, wind effect

Wind effect on buildings

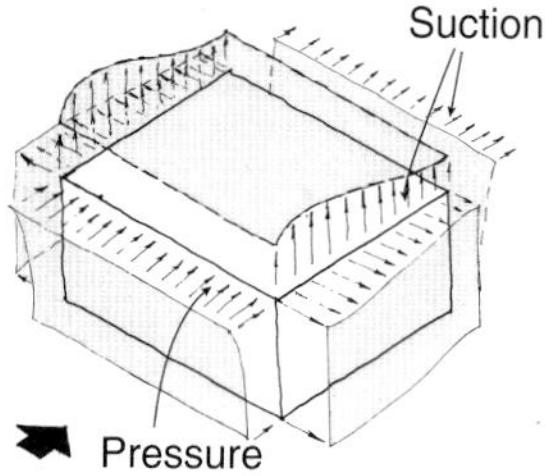

Wind effect to roof shape

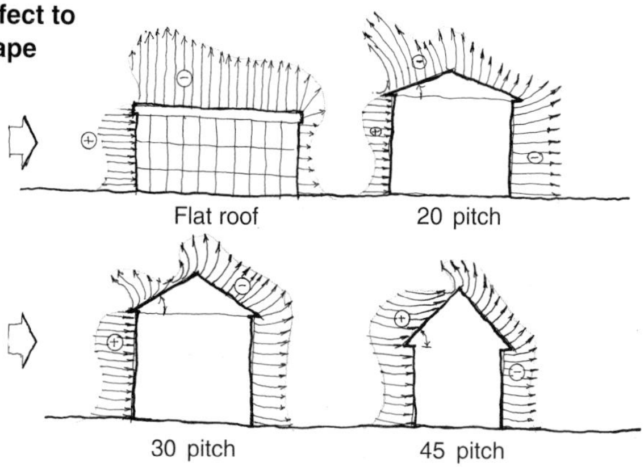

Extreme Weather: lightning

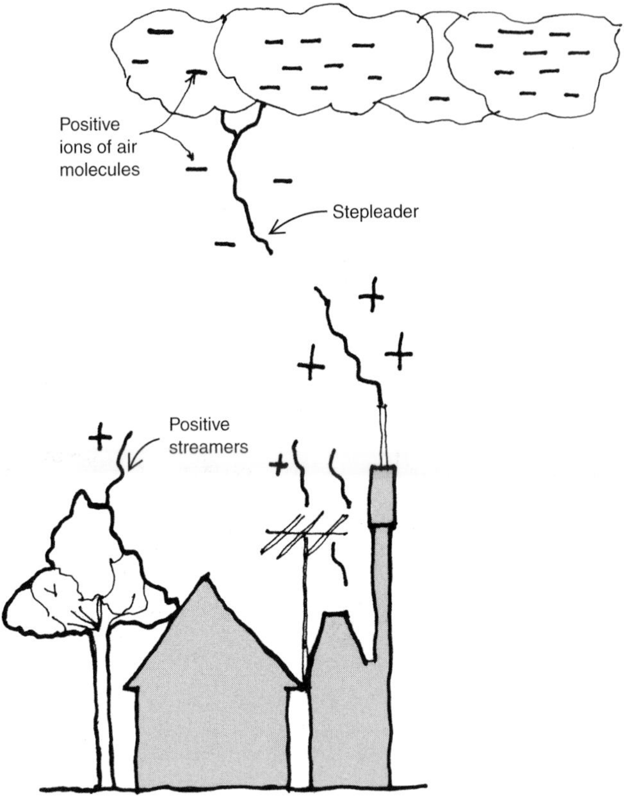
Positive
ions of air
molecules
Stepleader
Positive
streamers

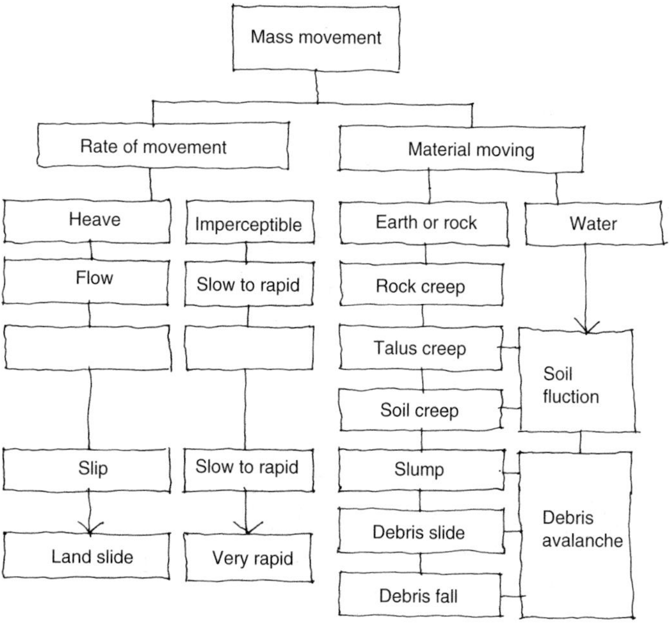
Mass movement
Rate of movement
Material moving
Heave
Imperceptible
Earth or rock
Water
Flow
Slow to rapid
Rock creep
Talus creep
Soil
fluction
Soil creep
Slip
Slow to rapid
Slump
Debris
avalanche
Debris slide
Land slide
Very rapid
Debris fall

Slide

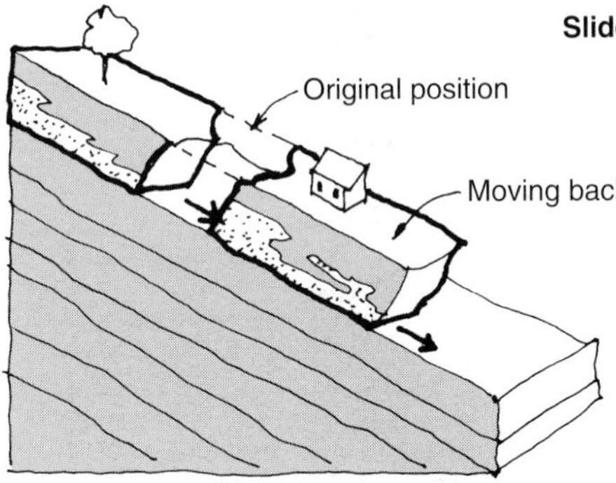
Original position
Moving back

Creep

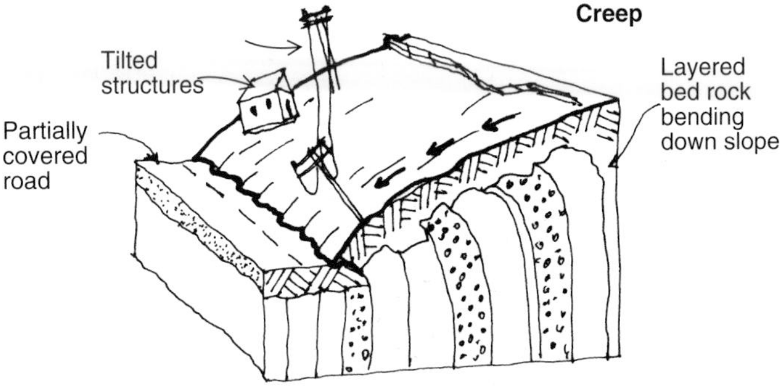
Tilted structures
Partially covered road
Layered bed rock bending down slope

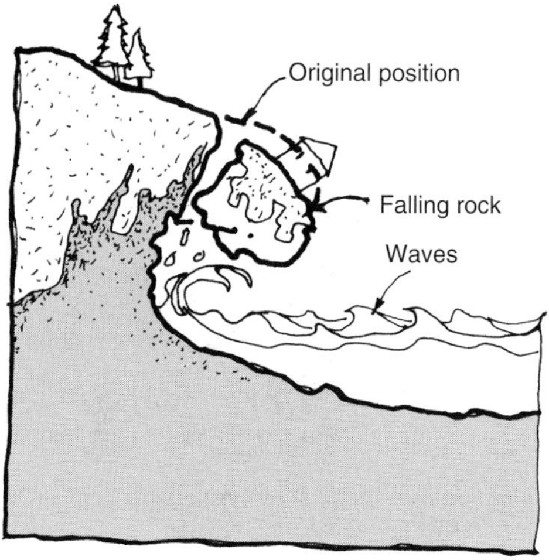
Original position
Falling rock
Waves

Fall

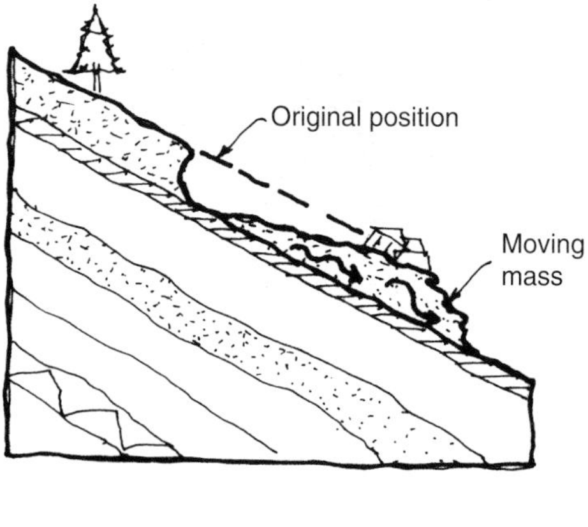
Original position
Moving
mass

Flow

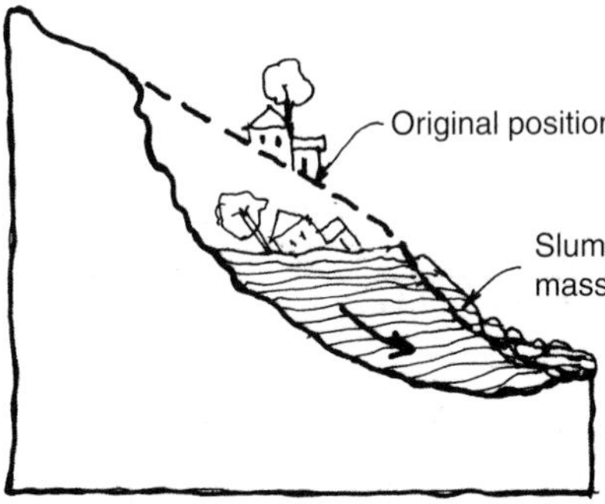
Original position
Slumped mass

Slump

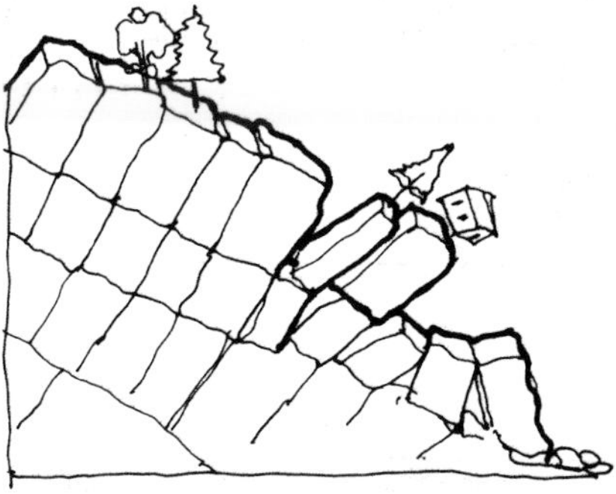

Topple

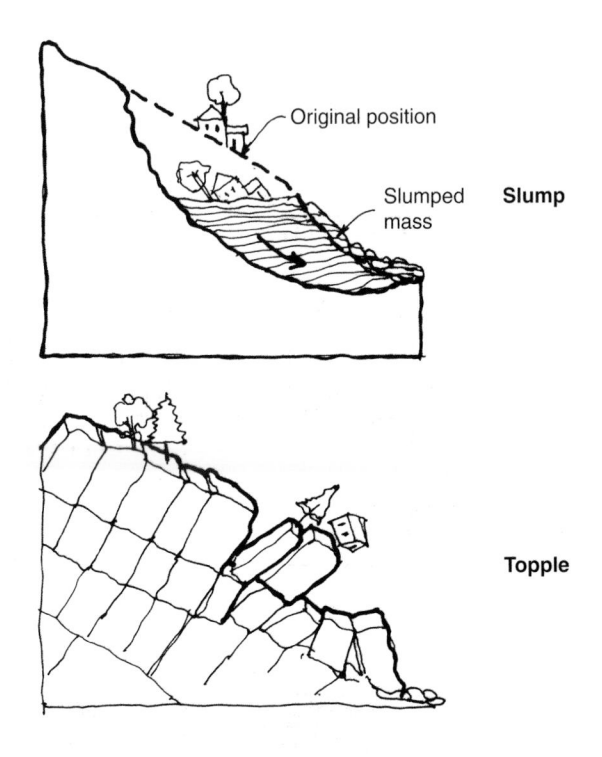

Heave effect

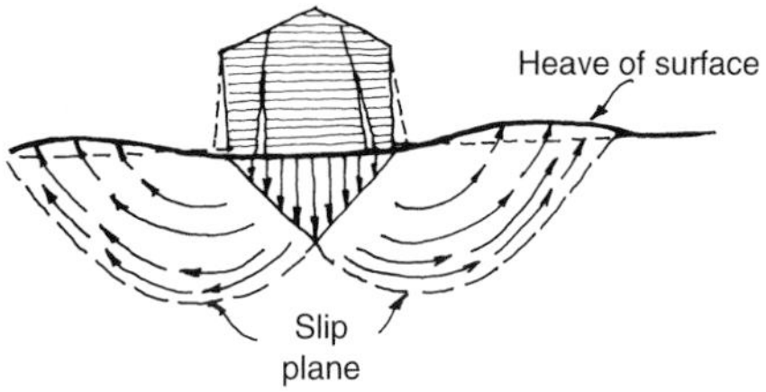

Instability of uneven ground effect on buildings

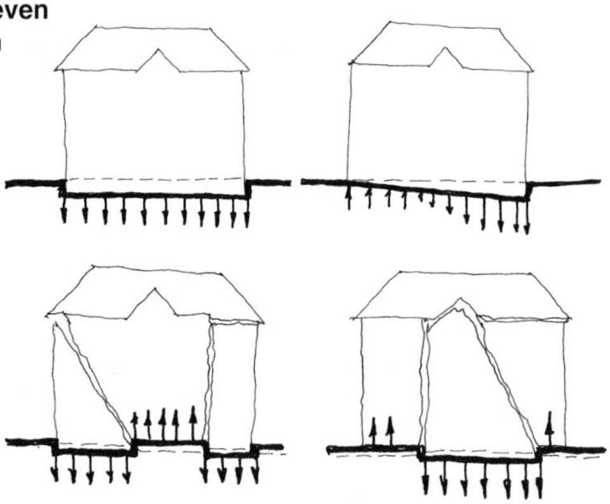

Typical structural damage due to ground effects

Flood

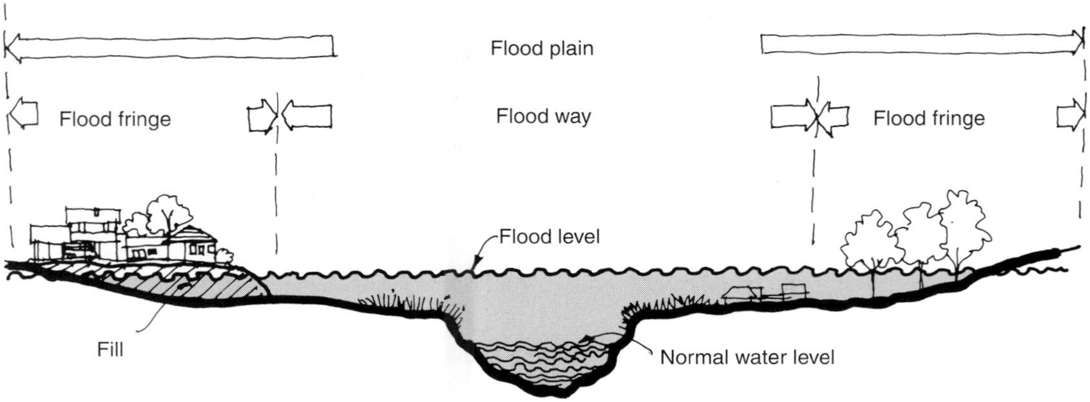
Flood plain
Flood fringe
Flood way
Flood fringe
Flood level
Fill
Normal water level

Volcano eruption

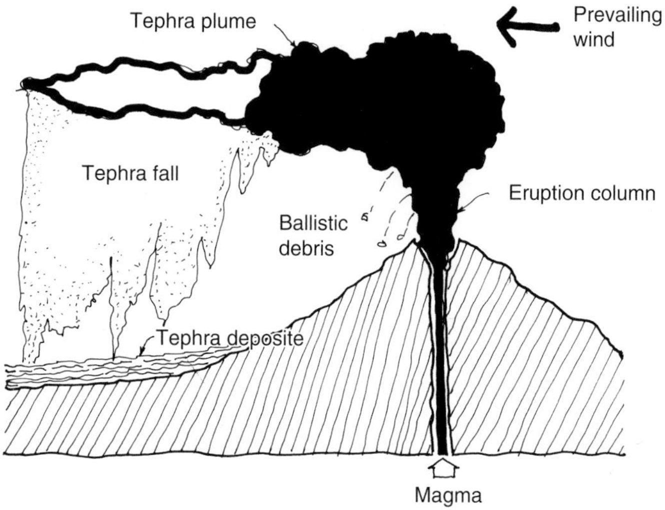
Tephra plume
Prevailing wind
Tephra fall
Eruption column
Ballistic debris
Tephra deposite
Magma

References

Architect's Data, Ernst Neufert, Crosby Lockwood Staples (1970).

Architect's Legal Handbook, Anthony Speaight and Gregory Stone, Architectural Press (1998).

Building Construction Vols I,II,III and IV, W.B. McKay, Longmans (1995).

The Building Design Easy Brief, Henry Haverstock, Morgan Grampain (1987).

Building Construction Handbook, R. Chudley, Laxton's (1988).

The Care and Conservation of Georgian Houses, Architectural Press with Edinburgh New Town Conservation Committee, Paul Harris Publishing (1978). Architectural Press (1980).

Dicobat – Editions Arcature (1990).

Dictionaire – Librarie Larousse (1981).

Drawing Office Practice for British Standard 1192, Architects and Builders (1953).

Ecohouse a design guide – Sue Roaf, Architectural Press (2002).

English Historic Carpentry, Cecil A. Hewett, Phillimore (1980).

Farms in England, Peter Fowler, Royal Commission on Historic Monuments, HMSO (1983).

Handbook of Urban Landscape, Cliff Tandy, Architectural Press (1975).

History of the English House, Nathaniel Lloyd, Architectural Press (1975).

Mitchell's Building Series,

Structure and Fabric 1, Jack Stroud Foster (1973).

Structure and Fabric 2, Jack Stroud Foster and Raymond Harrington (1976).

Components, Harold King (1983) Batsford Academic and Education.

Modern Practical Masonry, E.G. Warland, Sir Isaac Pitman & Sons Ltd., 2nd edn (1953).

Modulor Le Corbusier, Faber & Faber (1951).

New Metric Handbook, Edited by P. Tutt and D. Adler, Architectural Press (1979).

The Parish Churches of England, Charles Cox, B.T. Batsford (1954).

The Penguin Dictionary of Building, John S. Scott, Penguin (1982).

Repair Manual Reader's Digest (1976).

Sewage solutions – Nick Grant, Mark Moodie, Chriss Weedon – Centre for Alternative Technology Publications (2000).

References

Specification 1 – 6 Architectural Press (1987).

Sustainable Architecture – Brian Edwards, Architectural Press (1999).

Traditional Farm Buildings Richard Harris, Arts Council Exhibition Catalogue (1982).

Index

Index